RILEM State-of-the-Art Reports

Volume 48

RILEM, The International Union of Laboratories and Experts in Construction Materials, Systems and Structures, founded in 1947, is a non-governmental scientific association whose goal is to contribute to progress in the construction sciences, techniques and industries, essentially by means of the communication it fosters between research and practice. RILEM's focus is on construction materials and their use in building and civil engineering structures, covering all phases of the building process from manufacture to use and recycling of materials. More information on RILEM and its previous publications can be found on www.RILEM.net. The RILEM State-of-the-Art Reports (STAR) are produced by the Technical Committees. They represent one of the most important outputs that RILEM generates – high level scientific and engineering reports that provide cutting edge knowledge in a given field. The work of the TCs is one of RILEM's key functions. Members of a TC are experts in their field and give their time freely to share their expertise. As a result, the broader scientific community benefits greatly from RILEM's activities. RILEM's stated objective is to disseminate this information as widely as possible to the scientific community. RILEM therefore considers the STAR reports of its TCs as of highest importance, and encourages their publication whenever possible. The information in this and similar reports is mostly pre-normative in the sense that it provides the underlying scientific fundamentals on which standards and codes of practice are based. Without such a solid scientific basis, construction practice will be less than efficient or economical. It is RILEM's hope that this information will be of wide use to the scientific community.

Indexed in SCOPUS, Google Scholar and SpringerLink.

Véronique Baroghel-Bouny
Esperanza Menéndez Méndez • Manu Santhanam
Editors

External Sulfate Attacks - Mechanisms, Field Behavior and Lab Test Methods

State-of-the-Art Report of the RILEM Technical Committee 251-SRT

Springer

Editors
Véronique Baroghel-Bouny
University Gustave Eiffel
Marne-la-Vallée, France

Esperanza Menéndez Méndez
IETcc-CSIC
Madrid, Spain

Manu Santhanam
Indian Institute of Technology Madras
Chennai, India

ISSN 2213-204X ISSN 2213-2031 (electronic)
RILEM State-of-the-Art Reports
ISBN 978-3-032-19401-5 ISBN 978-3-032-19402-2 (eBook)
https://doi.org/10.1007/978-3-032-19402-2

This Springer imprint is published by the registered company Springer Nature Switzerland AG
The registered company address is: Gewerbestrasse 11, 6330 Cham, Switzerland

In Memory of Esperanza Menéndez Méndez

Organization RILEM TC 251-SRT

TC Chair: Dr Véronique BAROGHEL-BOUNY
Deputy chair: Dr Esperanza MENÉNDEZ MÉNDEZ
TC opened in **2013** and closed in **2019**
TC 251-SRT Members *(alphabetic order)*

Dr	ANDRADE	Carmen	Spain	CIMNE
Dr	BAROGHEL-BOUNY	Véronique	France	University Gustave Eiffel
Prof	CARCASSES	Myriam	France	LMDC—INSA/UPS Dept Genie Civil, Toulouse
Dr	CASSAGNABERE	Franck	France	LMDC—INSA/UPS Dept Genie Civil, Toulouse
Prof	De BELIE	Nele	Belgium	Ghent University
Dr	FERRARIS	Chiara "Clarissa"	USA	retired (former NIST)
	GARCÍA-ROVES LOZA	Ricardo	Spain	Universidad Politécnica de Madrid
Prof	GUO	Li-Ping	China	School of Materials Science & Engineering Southeast University Nanjing
Dr	HAERDTL	Reiner	Germany	Heidelberg Materials AG
	HAUFE	Johannes	Germany	Zert plus (former RWTH Aachen University)
Prof	HOOTON	Doug	Canada	University of Toronto
Dr	IZORET	Laurent	France	Réseau Réciproque Conseil (former France Ciment)
Dr	LEEMANN	Andreas	Switzerland	EMPA
Dr	LIPUS	Klaus	Germany	VDZ Technology gGmbH
Prof	LOUKILI	Ahmed	France	GeM—Ecole Centrale de Nantes
	MAES	Mathias	Belgium	SANACON (former Ghent University)
Dr	MARTINS	Isabel	Portugal	LNEC

Dr	MENÉNDEZ MÉNDEZ	Esperanza	Spain	IETcc (CSIC)
	MEULENYZER	Samuel	France	Holcim Innovation Center
Dr	MITTERMAYR	Florian	Austria	University of Innsbruck (former TU Graz)
Dr	OLONADE	Kolawole A	Nigeria	University of Lagos
Dr	OMIKRINE-METALSSI	Othman	France	University Gustave Eiffel
Dr	PRINCIGALLO	Antonio	Italy	(Italcementi) Heidelberg Materials, Italy
Prof	SANTHANAM	Manu	India	IITM (Indian Institute of Technology Madras)
Prof	SCRIVENER	Karen	Switzerland	EPFL - Laboratory of Construction Materials
Prof	SHI	Caijun	China	China Building Materials Academy
Dr	ŠTEFANCIC/ MLADENOVIC	Mateja /Ana	Slovenia	ZAG (Slovenian Nat Build & Civ Eng Institute)
Dr	TEGGUER DJERBI	Assia	France	University Gustave Eiffel
Dr	VILLAGRAN-ZACCARDI	Yury Andres	Belgium	VITO nv (Flemish Institute for Technological Research)
Dr	VOLLPRACHT	Anya	Germany	RWTH - Aachen University
Dr	WAN	Kai Tai	UK	Brunel University London
Dr	ZEGA	Claudio	Argentina	CONICET LEMIT La Plata

Other contributors to the STAR *(alphabetic order)*

Dr	AGUAYO	Fred	USA	University of Washington
Prof	GLASSER	Fredrik P.	UK	University of Aberdeen
Dr	HOSSACK	Ashlee	Canada	University of New Brunswick
Prof	ROZIERE	Emmanuel	France	GeM—Ecole Centrale de Nantes

RILEM Publications

The following list is presenting the global offer of RILEM Publications, sorted by series. Each publication is available in printed version and/or in online version.

RILEM Proceedings (PRO)

PRO 1: Durability of High Performance Concrete (ISBN: 2-912143-03-9; e-ISBN: 2-351580-12-5; e-ISBN: 2351580125); *Ed. H. Sommer*

PRO 2: Chloride Penetration into Concrete (ISBN: 2-912143-00-04; e-ISBN: 2912143454); *Eds. L.-O. Nilsson and J.-P. Ollivier*

PRO 3: Evaluation and Strengthening of Existing Masonry Structures (ISBN: 2-912143-02-0; e-ISBN: 2351580141); *Eds. L. Binda and C. Modena*

PRO 4: Concrete: From Material to Structure (ISBN: 2-912143-04-7; e-ISBN: 2351580206); *Eds. J.-P. Bournazel and Y. Malier*

PRO 5: The Role of Admixtures in High Performance Concrete (ISBN: 2-912143-05-5; e-ISBN: 2351580214); *Eds. J. G. Cabrera and R. Rivera-Villarreal*

PRO 6: High Performance Fiber Reinforced Cement Composites—HPFRCC 3 (ISBN: 2-912143-06-3; e-ISBN: 2351580222); *Eds. H. W. Reinhardt and A. E. Naaman*

PRO 7: 1st International RILEM Symposium on Self-Compacting Concrete (ISBN: 2-912143-09-8; e-ISBN: 2912143721); *Eds. Å. Skarendahl and Ö. Petersson*

PRO 8: International RILEM Symposium on Timber Engineering (ISBN: 2-912143-10-1; e-ISBN: 2351580230); *Ed. L. Boström*

PRO 9: 2nd International RILEM Symposium on Adhesion between Polymers and Concrete ISAP'99 (ISBN: 2-912143-11-X; e-ISBN: 2351580249); *Eds. Y. Ohama and M. Puterman*

PRO 10: 3rd International RILEM Symposium on Durability of Building and Construction Sealants (ISBN: 2-912143-13-6; e-ISBN: 2351580257); *Ed. A. T. Wolf*

PRO 11: 4th International RILEM Conference on Reflective Cracking in Pavements (ISBN: 2-912143-14-4; e-ISBN: 2351580265); *Eds. A. O. Abd El Halim, D. A. Taylor and El H. H. Mohamed*

PRO 12: International RILEM Workshop on Historic Mortars: Characteristics and Tests (ISBN: 2-912143-15-2; e-ISBN: 2351580273); *Eds. P. Bartos, C. Groot and J. J. Hughes*

PRO 13: 2nd International RILEM Symposium on Hydration and Setting (ISBN: 2-912143-16-0; e-ISBN: 2351580281); *Ed. A. Nonat*

PRO 14: Integrated Life-Cycle Design of Materials and Structures—ILCDES 2000 (ISBN: 951-758-408-3; e-ISBN: 235158029X); (ISSN: 0356-9403); *Ed. S. Sarja*

PRO 15: Fifth RILEM Symposium on Fibre-Reinforced Concretes (FRC)—BEFIB'2000 (ISBN: 2-912143-18-7; e-ISBN: 291214373X); *Eds. P. Rossi and G. Chanvillard*

PRO 16: Life Prediction and Management of Concrete Structures (ISBN: 2-912143-19-5; e-ISBN: 2351580303); *Ed. D. Naus*

PRO 17: Shrinkage of Concrete—Shrinkage 2000 (ISBN: 2-912143-20-9; e-ISBN: 2351580311); *Eds. V. Baroghel-Bouny and P.-C. Aïtcin*

PRO 18: Measurement and Interpretation of the On-Site Corrosion Rate (ISBN: 2-912143-21-7; e-ISBN: 235158032X); *Eds. C. Andrade, C. Alonso, J. Fullea, J. Polimon and J. Rodriguez*

PRO 19: Testing and Modelling the Chloride Ingress into Concrete (ISBN: 2-912143-22-5; e-ISBN: 2351580338); *Eds. C. Andrade and J. Kropp*

PRO 20: 1st International RILEM Workshop on Microbial Impacts on Building Materials (CD 02) (e-ISBN 978-2-35158-013-4); *Ed. M. Ribas Silva*

PRO 21: International RILEM Symposium on Connections between Steel and Concrete (ISBN: 2-912143-25-X; e-ISBN: 2351580346); *Ed. R. Eligehausen*

PRO 22: International RILEM Symposium on Joints in Timber Structures (ISBN: 2-912143-28-4; e-ISBN: 2351580354); *Eds. S. Aicher and H.-W. Reinhardt*

PRO 23: International RILEM Conference on Early Age Cracking in Cementitious Systems (ISBN: 2-912143-29-2; e-ISBN: 2351580362); *Eds. K. Kovler and A. Bentur*

PRO 24: 2nd International RILEM Workshop on Frost Resistance of Concrete (ISBN: 2-912143-30-6; e-ISBN: 2351580370); *Eds. M. J. Setzer, R. Auberg and H.-J. Keck*

PRO 25: International RILEM Workshop on Frost Damage in Concrete (ISBN: 2-912143-31-4; e-ISBN: 2351580389); *Eds. D. J. Janssen, M. J. Setzer and M. B. Snyder*

PRO 26: International RILEM Workshop on On-Site Control and Evaluation of Masonry Structures (ISBN: 2-912143-34-9; e-ISBN: 2351580141); *Eds. L. Binda and R. C. de Vekey*

PRO 27: International RILEM Symposium on Building Joint Sealants (CD03; e-ISBN: 235158015X); *Ed. A. T. Wolf*

PRO 28: 6th International RILEM Symposium on Performance Testing and Evaluation of Bituminous Materials—PTEBM'03 (ISBN: 2-912143-35-7; e-ISBN: 978-2-912143-77-8); *Ed. M. N. Partl*

PRO 29: 2nd International RILEM Workshop on Life Prediction and Ageing Management of Concrete Structures (ISBN: 2-912143-36-5; e-ISBN: 2912143780); *Ed. D. J. Naus*

PRO 30: 4th International RILEM Workshop on High Performance Fiber Reinforced Cement Composites—HPFRCC 4 (ISBN: 2-912143-37-3; e-ISBN: 2912143799); *Eds. A. E. Naaman and H. W. Reinhardt*

PRO 31: International RILEM Workshop on Test and Design Methods for Steel Fibre Reinforced Concrete: Background and Experiences (ISBN: 2-912143-38-1; e-ISBN: 2351580168); *Eds. B. Schnütgen and L. Vandewalle*

PRO 32: International Conference on Advances in Concrete and Structures 2 vol. (ISBN (set): 2-912143-41-1; e-ISBN: 2351580176); *Eds. Ying-shu Yuan, Surendra P. Shah and Heng-lin Lü*

PRO 33: 3rd International Symposium on Self-Compacting Concrete (ISBN: 2-912143-42-X; e-ISBN: 2912143713); *Eds. Ó. Wallevik and I. Níelsson*

PRO 34: International RILEM Conference on Microbial Impact on Building Materials (ISBN: 2-912143-43-8; e-ISBN: 2351580184); *Ed. M. Ribas Silva*

PRO 35: International RILEM TC 186-ISA on Internal Sulfate Attack and Delayed Ettringite Formation (ISBN: 2-912143-44-6; e-ISBN: 2912143802); *Eds. K. Scrivener and J. Skalny*

PRO 36: International RILEM Symposium on Concrete Science and Engineering—A Tribute to Arnon Bentur (ISBN: 2-912143-46-2; e-ISBN: 2912143586); *Eds. K. Kovler, J. Marchand, S. Mindess and J. Weiss*

PRO 37: 5th International RILEM Conference on Cracking in Pavements—Mitigation, Risk Assessment and Prevention (ISBN: 2-912143-47-0; e-ISBN: 2912143764); *Eds. C. Petit, I. Al-Qadi and A. Millien*

PRO 38: 3rd International RILEM Workshop on Testing and Modelling the Chloride Ingress into Concrete (ISBN: 2-912143-48-9; e-ISBN: 2912143578); *Eds. C. Andrade and J. Kropp*

PRO 39: 6th International RILEM Symposium on Fibre-Reinforced Concretes—BEFIB 2004 (ISBN: 2-912143-51-9; e-ISBN: 2912143748); *Eds. M. Di Prisco, R. Felicetti and G. A. Plizzari*

PRO 40: International RILEM Conference on the Use of Recycled Materials in Buildings and Structures (ISBN: 2-912143-52-7; e-ISBN: 2912143756); *Eds. E. Vázquez, Ch. F. Hendriks and G. M. T. Janssen*

PRO 41: RILEM International Symposium on Environment-Conscious Mate- rials and Systems for Sustainable Development (ISBN: 2-912143-55-1; e-ISBN: 2912143640); *Eds. N. Kashino and Y. Ohama*

PRO 42: SCC'2005—China: 1st International Symposium on Design, Performance and Use of Self-Consolidating Concrete (ISBN: 2-912143-61-6; e-ISBN: 2912143624); *Eds. Zhiwu Yu, Caijun Shi, Kamal Henri Khayat and Youjun Xie*

PRO 43: International RILEM Workshop on Bonded Concrete Overlays (e-ISBN: 2-912143-83-7); *Eds. J. L. Granju and J. Silfwerbrand*

PRO 44: 2nd International RILEM Workshop on Microbial Impacts on Building Materials (CD11) (e-ISBN: 2-912143-84-5); *Ed. M. Ribas Silva*

PRO 45: 2nd International Symposium on Nanotechnology in Construction, Bilbao (ISBN: 2-912143-87-X; e-ISBN: 2912143888); *Eds. Peter J. M. Bartos, Yolanda de Miguel and Antonio Porro*

PRO 46: ConcreteLife'06—International RILEM-JCI Seminar on Concrete Durability and Service Life Planning: Curing, Crack Control, Performance in Harsh Environments (ISBN: 2-912143-89-6; e-ISBN: 291214390X); *Ed. K. Kovler*

PRO 47: International RILEM Workshop on Performance Based Evaluation and Indicators for Concrete Durability (ISBN: 978-2-912143-95-2; e-ISBN: 9782912143969); *Eds. V. Baroghel-Bouny, C. Andrade, R. Torrent and K. Scrivener*

PRO 48: 1st International RILEM Symposium on Advances in Concrete through Science and Engineering (e-ISBN: 2-912143-92-6); *Eds. J. Weiss, K. Kovler, J. Marchand, and S. Mindess*

PRO 49: International RILEM Workshop on High Performance Fiber Rein- forced Cementitious Composites in Structural Applications (ISBN: 2-912143-93-4; e-ISBN: 2912143942); *Eds. G. Fischer and V. C. Li*

PRO 50: 1st International RILEM Symposium on Textile Reinforced Concrete (ISBN: 2-912143-97-7; e-ISBN: 2351580087); *Eds. Josef Hegger, Wolfgang Brameshuber and Norbert Will*

PRO 51: 2nd International Symposium on Advances in Concrete through Science and Engineering (ISBN: 2-35158-003-6; e-ISBN: 2-35158-002-8); *Eds. J. Marchand, B. Bissonnette, R. Gagné, M. Jolin and F. Paradis*

PRO 52: Volume Changes of Hardening Concrete: Testing and Mitigation (ISBN: 2-35158-004-4; e-ISBN: 2-35158-005-2); *Eds. O. M. Jensen, P. Lura and K. Kovler*

PRO 53: High Performance Fiber Reinforced Cement Composites—HPFRCC5 (ISBN: 978-2-35158-046-2; e-ISBN: 978-2-35158-089-9); *Eds. H. W. Reinhardt and A. E. Naaman*

PRO 54: 5th International RILEM Symposium on Self-Compacting Concrete (ISBN: 978-2-35158-047-9; e-ISBN: 978-2-35158-088-2); *Eds. G. De Schutter and V. Boel*

PRO 55: International RILEM Symposium Photocatalysis, Environment and ConstructionMaterials(ISBN:978-2-35158-056-1;e-ISBN:978-2-35158-057-8); *Eds. P. Baglioni and L. Cassar*

PRO 56: International RILEM Workshop on Integral Service Life Modelling of Concrete Structures (ISBN 978-2-35158-058-5; e-ISBN: 978-2-35158-090-5); *Eds. R. M. Ferreira, J. Gulikers and C. Andrade*

PRO 57: RILEM Workshop on Performance of cement-based materials in aggressive aqueous environments (e-ISBN: 978-2-35158-059-2); *Ed. N. De Belie*

PRO 58: International RILEM Symposium on Concrete Modelling—CONMOD'08 (ISBN: 978-2-35158-060-8; e-ISBN: 978-2-35158-076-9); *Eds. E. Schlangen and G. De Schutter*

PRO 59: International RILEM Conference on On Site Assessment of Concrete, Masonry and Timber Structures—SACoMaTiS 2008 (ISBN set: 978-2-35158-061-5; e-ISBN: 978-2-35158-075-2); *Eds. L. Binda, M. di Prisco and R. Felicetti*

PRO 60: Seventh RILEM International Symposium on Fibre Reinforced Concrete: Design and Applications—BEFIB 2008 (ISBN: 978-2-35158-064-6; e-ISBN: 978-2-35158-086-8); *Ed. R. Gettu*

PRO 61: 1st International Conference on Microstructure Related Durability of Cementitious Composites 2 vol., (ISBN: 978-2-35158-065-3; e-ISBN: 978-2-35158-084-4); *Eds. W. Sun, K. van Breugel, C. Miao, G. Ye and H. Chen*

PRO 62: NSF/RILEM Workshop: In-situ Evaluation of Historic Wood and Masonry Structures (e-ISBN: 978-2-35158-068-4); *Eds. B. Kasal, R. Anthony and M. Drdácký*

PRO 63: Concrete in Aggressive Aqueous Environments: Performance, Testing and Modelling, 2 vol., (ISBN: 978-2-35158-071-4; e-ISBN: 978-2-35158-082-0); *Eds. M. G. Alexander and A. Bertron*

PRO 64: Long Term Performance of Cementitious Barriers and Reinforced Concrete in Nuclear Power Plants and Waste Management—NUCPERF 2009 (ISBN: 978-2-35158-072-1; e-ISBN: 978-2-35158-087-5); *Eds. V. L'Hostis, R. Gens and C. Gallé*

PRO 65: Design Performance and Use of Self-consolidating Concrete—SCC'2009 (ISBN: 978-2-35158-073-8; e-ISBN: 978-2-35158-093-6); *Eds. C. Shi, Z. Yu, K. H. Khayat and P. Yan*

PRO 66: 2nd International RILEM Workshop on Concrete Durability and Service Life Planning—ConcreteLife'09 (ISBN: 978-2-35158-074-5; ISBN: 978-2-35158-074-5); *Ed. K. Kovler*

PRO 67: Repairs Mortars for Historic Masonry (e-ISBN: 978-2-35158-083-7); *Ed. C. Groot*

PRO 68: Proceedings of the 3rd International RILEM Symposium on 'Rheology of Cement Suspensions such as Fresh Concrete (ISBN 978-2-35158-091-2; e-ISBN: 978-2-35158-092-9); *Eds. O. H. Wallevik, S. Kubens and S. Oesterheld*

PRO 69: 3rd International PhD Student Workshop on 'Modelling the Durability of Reinforced Concrete (ISBN: 978-2-35158-095-0); *Eds. R. M. Ferreira, J. Gulikers and C. Andrade*

PRO 70: 2nd International Conference on 'Service Life Design for Infrastructure' (ISBN set: 978-2-35158-096-7, e-ISBN: 978-2-35158-097-4); *Eds. K. van Breugel, G. Ye and Y. Yuan*

PRO 71: Advances in Civil Engineering Materials—The 50-year Teaching Anniversary of Prof. Sun Wei' (ISBN: 978-2-35158-098-1; e-ISBN: 978-2-35158-099-8); *Eds. C. Miao, G. Ye and H. Chen*

PRO 72: First International Conference on 'Advances in Chemically-Activated Materials—CAM'2010' (2010), 264 pp., ISBN: 978-2-35158-101-8; e-ISBN: 978-2-35158-115-5; *Eds. Caijun Shi and Xiaodong Shen*

PRO 73: 2nd International Conference on 'Waste Engineering and Management—ICWEM 2010' (2010), 894 pp., ISBN: 978-2-35158-102-5; e-ISBN: 978-2-35158-103-2, *Eds. J. Zh. Xiao, Y. Zhang, M. S. Cheung and R. Chu*

PRO 74: International RILEM Conference on 'Use of Superabsorbent Polymers and Other New Additives in Concrete' (2010) 374 pp., ISBN: 978-2-35158-104-9; e-ISBN: 978-2-35158-105-6; *Eds. O. M. Jensen, M. T. Hasholt, and S. Laustsen*

PRO 75: International Conference on 'Material Science—2nd ICTRC—Textile Reinforced Concrete—Theme 1' (2010) 436 pp., ISBN: 978-2-35158-106-3; e-ISBN: 978-2-35158-107-0; *Ed. W. Brameshuber*

PRO 76: International Conference on 'Material Science—HetMat—Modelling of Heterogeneous Materials—Theme 2' (2010) 255 pp., ISBN: 978-2-35158-108-7; e-ISBN: 978-2-35158-109-4; *Ed. W. Brameshuber*

PRO 77: International Conference on 'Material Science—AdIPoC—Additions Improving Properties of Concrete—Theme 3' (2010) 459 pp., ISBN: 978-2-35158-110-0; e-ISBN: 978-2-35158-111-7; *Ed. W. Brameshuber*

PRO 78: 2nd Historic Mortars Conference and RILEM TC 203-RHM Final Workshop—HMC2010 (2010) 1416 pp., e-ISBN: 978-2-35158-112-4; *Eds. J. Válek, C. Groot and J. J. Hughes*

PRO 79: International RILEM Conference on Advances in Construction Materials Through Science and Engineering (2011) 213 pp., ISBN: 978-2-35158-116-2, e- ISBN: 978-2-35158-117-9; *Eds. Christopher Leung and K. T. Wan*

PRO 80: 2nd International RILEM Conference on Concrete Spalling due to Fire Exposure(2011)453pp.,ISBN:978-2-35158-118-6;e-ISBN:978-2-35158-119-3; *Eds. E.A.B. Koenders and F. Dehn*

PRO 81: 2nd International RILEM Conference on Strain Hardening Cementitious Composites (SHCC2-Rio) (2011) 451 pp., ISBN: 978-2-35158-120-9; e-ISBN: 978-2-35158-121-6; *Eds. R.D. Toledo Filho, F.A. Silva, E.A.B. Koenders and E.M.R. Fairbairn*

PRO 82: 2nd International RILEM Conference on Progress of Recycling in the Built Environment (2011) 507 pp., e-ISBN: 978-2-35158-122-3; *Eds. V.M. John, E. Vazquez, S.C. Angulo and C. Ulsen*

PRO 83: 2nd International Conference on Microstructural-related Durability of Cementitious Composites (2012) 250 pp., ISBN: 978-2-35158-129-2; e-ISBN: 978-2-35158-123-0; *Eds. G. Ye, K. van Breugel, W. Sun and C. Miao*

PRO 84: CONSEC13—Seventh International Conference on Concrete under Severe Conditions—Environment and Loading (2013) 1930 pp., ISBN: 978-2-35158-124-7; e-ISBN: 978-2-35158-134-6; *Eds. Z.J. Li, W. Sun, C.W. Miao, K. Sakai, O.E. Gjorv and N. Banthia*

PRO 85: RILEM-JCI International Workshop on Crack Control of Mass Concrete and Related issues concerning Early-Age of Concrete Structures—ConCrack 3— Control of Cracking in Concrete Structures 3 (2012) 237 pp., ISBN: 978-2-35158-125-4; e-ISBN: 978-2-35158-126-1; *Eds. F. Toutlemonde and J.-M. Torrenti*

PRO 86: International Symposium on Life Cycle Assessment and Construction (2012) 414 pp., ISBN: 978-2-35158-127-8, e-ISBN: 978-2-35158-128-5; *Eds. A. Ventura and C. de la Roche*

PRO 87: UHPFRC 2013—RILEM-fib-AFGC International Symposium on Ultra-High Performance Fibre-Reinforced Concrete (2013), ISBN: 978-2-35158-130-8, e-ISBN: 978-2-35158-131-5; *Eds. F. Toutlemonde*

PRO 88: 8th RILEM International Symposium on Fibre Reinforced Concrete (2012) 344 pp., ISBN: 978-2-35158-132-2; e-ISBN: 978-2-35158-133-9; *Eds. Joaquim A.O. Barros*

PRO 89: RILEM International workshop on performance-based specification and control of concrete durability (2014) 678 pp., ISBN: 978-2-35158-135-3; e-ISBN: 978-2-35158-136-0; *Eds. D. Bjegović, H. Beushausen and M. Serdar*

PRO 90: 7th RILEM International Conference on Self-Compacting Concrete and of the 1st RILEM International Conference on Rheology and Processing of Construction Materials (2013) 396 pp., ISBN: 978-2-35158-137-7; e-ISBN: 978-2-35158-138-4; *Eds. Nicolas Roussel and Hela Bessaies-Bey*

PRO 91: CONMOD 2014—RILEM International Symposium on Concrete Modelling (2014), ISBN: 978-2-35158-139-1; e-ISBN: 978-2-35158-140-7; *Eds. Kefei Li, Peiyu Yan and Rongwei Yang*

PRO 92: CAM 2014—2nd International Conference on advances in chemically-activated materials (2014) 392 pp., ISBN: 978-2-35158-141-4; e-ISBN: 978-2-35158-142-1; *Eds. Caijun Shi and Xiadong Shen*

PRO 93: SCC 2014—3rd International Symposium on Design, Performance and Use of Self-Consolidating Concrete (2014) 438 pp., ISBN: 978-2-35158-143-8; e-ISBN: 978-2-35158-144-5; *Eds. Caijun Shi, Zhihua Ou and Kamal H. Khayat*

PRO 94 (online version): HPFRCC-7—7th RILEM conference on High performance fiber reinforced cement composites (2015), e-ISBN: 978-2-35158-146-9; *Eds. H.W. Reinhardt, G.J. Parra-Montesinos and H. Garrecht*

PRO 95: International RILEM Conference on Application of superabsorbent polymers and other new admixtures in concrete construction (2014), ISBN: 978-2-35158-147-6; e-ISBN: 978-2-35158-148-3; *Eds. Viktor Mechtcherine and Christof Schroefl*

PRO 96 (online version): XIII DBMC: XIII International Conference on Durability of Building Materials and Components (2015), e-ISBN: 978-2-35158-149-0; *Eds. M. Quattrone and V.M. John*

PRO 97: SHCC3—3rd International RILEM Conference on Strain Hardening Cementitious Composites (2014), ISBN: 978-2-35158-150-6; e-ISBN: 978-2-35158-151-3; *Eds. E. Schlangen, M.G. Sierra Beltran, M. Lukovic and G. Ye*

PRO 98: FERRO-11—11th International Symposium on Ferrocement and 3rd ICTRC—International Conference on Textile Reinforced Concrete (2015), ISBN: 978-2-35158-152-0; e-ISBN: 978-2-35158-153-7; *Ed. W. Brameshuber*

PRO 99 (online version): ICBBM 2015—1st International Conference on Bio-Based Building Materials (2015), e-ISBN: 978-2-35158-154-4; *Eds. S. Amziane and M. Sonebi*

PRO 100: SCC16—RILEM Self-Consolidating Concrete Conference (2016), ISBN: 978-2-35158-156-8; e-ISBN: 978-2-35158-157-5; *Ed. Kamal H. Kayat*

PRO 101 (online version): III Progress of Recycling in the Built Environment (2015), e-ISBN: 978-2-35158-158-2; *Eds I. Martins, C. Ulsen and S. C. Angulo*

PRO 102 (online version): RILEM Conference on Microorganisms-Cementitious Materials Interactions (2016), e-ISBN: 978-2-35158-160-5; *Eds. Alexandra Bertron, Henk Jonkers and Virginie Wiktor*

PRO 103 (online version): ACESC'16—Advances in Civil Engineering and Sustainable Construction (2016), e-ISBN: 978-2-35158-161-2; *Eds. T.Ch. Madhavi, G. Prabhakar, Santhosh Ram and P.M. Rameshwaran*

PRO 104 (online version): SSCS'2015—Numerical Modeling—Strategies for Sustainable Concrete Structures (2015), e-ISBN: 978-2-35158-162-9

PRO 105: 1st International Conference on UHPC Materials and Structures (2016), ISBN: 978-2-35158-164-3; e-ISBN: 978-2-35158-165-0

PRO 106: AFGC-ACI-fib-RILEM International Conference on Ultra-High-Performance Fibre-Reinforced Concrete—UHPFRC 2017 (2017), ISBN: 978-2-35158-166-7; e-ISBN: 978-2-35158-167-4; *Eds. François Toutlemonde and Jacques Resplendino*

PRO 107 (online version): XIV DBMC—14th International Conference on Durability of Building Materials and Components (2017), e-ISBN: 978-2-35158-159-9; *Eds. Geert De Schutter, Nele De Belie, Arnold Janssens and Nathan Van Den Bossche*

PRO 108: MSSCE 2016—Innovation of Teaching in Materials and Structures (2016), ISBN: 978-2-35158-178-0; e-ISBN: 978-2-35158-179-7; *Ed. Per Golter-mann*

PRO 109 (2 volumes): MSSCE 2016—Service Life of Cement-Based Materials and Structures (2016), ISBN Vol. 1: 978-2-35158-170-4; Vol. 2: 978-2-35158-171-4; Set Vol. 1&2: 978-2-35158-172-8; e-ISBN: 978-2-35158-173-5; *Eds. Miguel Azenha, Ivan Gabrijel, Dirk Schlicke, Terje Kanstad and Ole Mejlhede Jensen*

PRO 110: MSSCE 2016—Historical Masonry (2016), ISBN: 978-2-35158-178-0; e-ISBN: 978-2-35158-179-7; *Eds. Inge Rörig-Dalgaard and Ioannis Ioannou*

PRO 111: MSSCE 2016—Electrochemistry in Civil Engineering (2016); ISBN: 978-2-35158-176-6; e-ISBN: 978-2-35158-177-3; *Ed. Lisbeth M. Ottosen*

PRO 112: MSSCE 2016—Moisture in Materials and Structures (2016), ISBN: 978-2-35158-178-0; e-ISBN: 978-2-35158-179-7; *Eds. Kurt Kielsgaard Hansen, Carsten Rode and Lars-Olof Nilsson*

PRO 113: MSSCE 2016—Concrete with Supplementary Cementitious Materials (2016), ISBN: 978-2-35158-178-0; e-ISBN: 978-2-35158-179-7; *Eds. Ole Mejlhede Jensen, Konstantin Kovler and Nele De Belie*

PRO 114: MSSCE 2016—Frost Action in Concrete (2016), ISBN: 978-2-35158-182-7; e-ISBN: 978-2-35158-183-4; *Eds. Marianne Tange Hasholt, Katja Fridh and R. Doug Hooton*

PRO 115: MSSCE 2016—Fresh Concrete (2016), ISBN: 978-2-35158-184-1; e-ISBN: 978-2-35158-185-8; *Eds. Lars N. Thrane, Claus Pade, Oldrich Svec and Nicolas Roussel*

PRO 116: BEFIB 2016—9th RILEM International Symposium on Fiber Reinforced Concrete (2016), ISBN: 978-2-35158-187-2; e-ISBN: 978-2-35158-186-5; *Eds. N. Banthia, M. di Prisco and S. Soleimani-Dashtaki*

PRO 117: 3rd International RILEM Conference on Microstructure Related Durability of Cementitious Composites (2016), ISBN: 978-2-35158-188-9; e-ISBN: 978-2-35158-189-6; *Eds. Changwen Miao, Wei Sun, Jiaping Liu, Huisu Chen, Guang Ye and Klaas van Breugel*

PRO 118 (4 volumes): International Conference on Advances in Construction Materials and Systems (2017), ISBN Set: 978-2-35158-190-2; Vol. 1: 978-2-35158-193-3; Vol. 2: 978-2-35158-194-0; Vol. 3: ISBN: 978-2-35158-195-7; Vol. 4: ISBN: 978-2-35158-196-4; e-ISBN: 978-2-35158-191-9; *Ed. Manu Santhanam*

PRO 119 (online version): ICBBM 2017—Second International RILEM Conference on Bio-based Building Materials, (2017), e-ISBN: 978-2-35158-192-6; *Ed. Sofiane Amziane*

PRO 120 (2 volumes): EAC-02—2nd International RILEM/COST Conference on Early Age Cracking and Serviceability in Cement-based Materials and Structures, (2017), Vol. 1: 978-2-35158-199-5, Vol. 2: 978-2-35158-200-8, Set: 978-2-35158-197-1, e-ISBN: 978-2-35158-198-8; *Eds. Stéphanie Staquet and Dimitrios Aggelis*

PRO 121 (2 volumes): SynerCrete18: Interdisciplinary Approaches for Cement-based Materials and Structural Concrete: Synergizing Expertise and Bridging Scales of Space and Time, (2018), Set: 978-2-35158-202-2, Vol.1: 978-2-35158-211-4, Vol.2: 978-2-35158-212-1, e-ISBN: 978-2-35158-203-9; *Eds. Miguel Azenha, Dirk Schlicke, Farid Benboudjema, Agnieszka Knoppik*

PRO 122: SCC'2018 China—Fourth International Symposium on Design, Performance and Use of Self-Consolidating Concrete, (2018), ISBN: 978-2-35158-204-6, e-ISBN: 978-2-35158-205-3; *Eds. C. Shi, Z. Zhang, K. H. Khayat*

PRO 123: Final Conference of RILEM TC 253-MCI: Microorganisms-Cementitious Materials Interactions (2018), Set: 978-2-35158-207-7, Vol.1: 978-2-35158-209-1, Vol.2: 978-2-35158-210-7, e-ISBN: 978-2-35158-206-0; *Ed. Alexandra Bertron*

PRO 124 (online version): Fourth International Conference Progress of Recycling in the Built Environment (2018), e-ISBN: 978-2-35158-208-4; *Eds. Isabel M. Martins, Carina Ulsen, Yury Villagran*

PRO 125 (online version): SLD4—4th International Conference on Service Life Design for Infrastructures (2018), e-ISBN: 978-2-35158-213-8; *Eds. Guang Ye, Yong Yuan, Claudia Romero Rodriguez, Hongzhi Zhang, Branko Savija*

PRO 126: Workshop on Concrete Modelling and Material Behaviour in honor of Professor Klaas van Breugel (2018), ISBN: 978-2-35158-214-5, e-ISBN: 978-2-35158-215-2; *Ed. Guang Ye*

PRO 127 (online version): CONMOD2018—Symposium on Concrete Modelling (2018), e-ISBN: 978-2-35158-216-9; *Eds. Erik Schlangen, Geert de Schutter, Branko Savija, Hongzhi Zhang, Claudia Romero Rodriguez*

PRO 128: SMSS2019—International Conference on Sustainable Materials, Systems and Structures (2019), ISBN: 978-2-35158-217-6, e-ISBN: 978-2-35158-218-3

PRO 129: 2nd International Conference on UHPC Materials and Structures (UHPC2018-China), ISBN: 978-2-35158-219-0, e-ISBN: 978-2-35158-220-6

PRO 130: 5th Historic Mortars Conference (2019), ISBN: 978-2-35158-221-3, e-ISBN: 978-2-35158-222-0; *Eds. José Ignacio Álvarez, José María Fernández, Íñigo Navarro, Adrián Durán, Rafael Sirera*

PRO 131 (online version): 3rd International Conference on Bio-Based Building Materials (ICBBM2019), e-ISBN: 978-2-35158-229-9; *Eds. Mohammed Sonebi, Sofiane Amziane, Jonathan Page*

PRO 132: IRWRMC'18—International RILEM Workshop on Rheological Measure- ments of Cement-based Materials (2018), ISBN: 978-2-35158-230-5, e-ISBN: 978-2-35158-231-2; *Eds. Chafika Djelal, Yannick Vanhove*

PRO 133 (online version): CO2STO2019—International Workshop CO2 Storage in Concrete (2019), e-ISBN: 978-2-35158-232-9; *Eds. Assia Djerbi, Othman Omikrine- Metalssi, Teddy Fen-Chong*

PRO 134: 3rd ACF/HNU International Conference on UHPC Materials and Structures—UHPC'2020, ISBN: 978-2-35158-233-6, e-ISBN: 978-2-35158-234-3; *Eds. Caijun Shi and Jiaping Liu*

PRO 135: Fourth International Conference on Chemically Activated Materials (CAM2021), ISBN: 978-2-35158-235-0, e-ISBN: 978-2-35158-236-7; *Eds. Caijun Shi and Xiang Hu*

PRO 136: Lime Based Materials for Repairing Historic Structures (2022), ISBN: 978-2-35158-240-4, e-ISBN: 978-2-35158-237-4; *Eds. Ioanna Papayianni and Jan Válek*

PRO 137 (online version): V International Conference Progress of Recycling in the Built Environment (2023), e-ISBN: 978-2-35158-238-1; *Eds. Ines Döring, Yury Villagrán Zaccardi, Anette Müller*

RILEM Reports (REP)

Report 19: Considerations for Use in Managing the Aging of Nuclear Power Plant Concrete Structures (ISBN: 2-912143-07-1); *Ed. D. J. Naus*

Report 20: Engineering and Transport Properties of the Interfacial Transition Zone in Cementitious Composites (ISBN: 2-912143-08-X); *Eds. M. G. Alexander, G. Arliguie, G. Ballivy, A. Bentur and J. Marchand*

Report 21: Durability of Building Sealants (ISBN: 2-912143-12-8); *Ed. A. T. Wolf*

Report 22: Sustainable Raw Materials—Construction and Demolition Waste (ISBN: 2-912143-17-9); *Eds. C. F. Hendriks and H. S. Pietersen*

Report 23: Self-Compacting Concrete state-of-the-art report (ISBN: 2-912143-23-3); *Eds. Å. Skarendahl and Ö. Petersson*

Report 24: Workability and Rheology of Fresh Concrete: Compendium of Tests (ISBN: 2-912143-32-2); *Eds. P. J. M. Bartos, M. Sonebi and A. K. Tamimi*

Report 25: Early Age Cracking in Cementitious Systems (ISBN: 2-912143-33-0); *Ed. A. Bentur*

Report 26: Towards Sustainable Roofing (Joint Committee CIB/RILEM) (CD 07) (e-ISBN 978-2-912143-65-5); *Eds. Thomas W. Hutchinson and Keith Roberts*

Report 27: Condition Assessment of Roofs (Joint Committee CIB/RILEM) (CD 08) (e-ISBN 978-2-912143-66-2); *Ed. CIB W 83/RILEM TC166-RMS*

Report 28: Final report of RILEM TC 167-COM 'Characterisation of Old Mortars with Respect to Their Repair (ISBN: 978-2-912143-56-3); *Eds. C. Groot, G. Ashall and J. Hughes*

Report 29: Pavement Performance Prediction and Evaluation (PPPE): Interlaboratory Tests (e-ISBN: 2-912143-68-3); *Eds. M. Partl and H. Piber*

Report 30: Final Report of RILEM TC 198-URM 'Use of Recycled Materials' (ISBN: 2-912143-82-9; e-ISBN: 2-912143-69-1); *Eds. Ch. F. Hendriks, G. M. T. Janssen and E. Vázquez*

Report 31: Final Report of RILEM TC 185-ATC 'Advanced testing of cement-based materials during setting and hardening' (ISBN: 2-912143-81-0; e-ISBN: 2-912143-70-5); *Eds. H. W. Reinhardt and C. U. Grosse*

Report 32: Probabilistic Assessment of Existing Structures. A JCSS publication (ISBN 2-912143-24-1); *Ed. D. Diamantidis*

Report 33: State-of-the-Art Report of RILEM Technical Committee TC 184-IFE 'Industrial Floors' (ISBN 2-35158-006-0); *Ed. P. Seidler*

Report 34: Report of RILEM Technical Committee TC 147-FMB 'Fracture mechanics applications to anchorage and bond' Tension of Reinforced Concrete Prisms—Round Robin Analysis and Tests on Bond (e-ISBN 2-912143-91-8); *Eds. L. Elfgren and K. Noghabai*

Report 35: Final Report of RILEM Technical Committee TC 188-CSC 'Casting of Self Compacting Concrete' (ISBN 2-35158-001-X; e-ISBN: 2-912143-98-5); *Eds. Å. Skarendahl and P. Billberg*

Report 36: State-of-the-Art Report of RILEM Technical Committee TC 201-TRC 'Textile Reinforced Concrete' (ISBN 2-912143-99-3); *Ed. W. Brameshuber*

Report 37: State-of-the-Art Report of RILEM Technical Committee TC 192-ECM 'Environment-conscious construction materials and systems' (ISBN: 978-2-35158-053-0); *Eds. N. Kashino, D. Van Gemert and K. Imamoto*

Report 38: State-of-the-Art Report of RILEM Technical Committee TC 205-DSC 'Durability of Self-Compacting Concrete' (ISBN: 978-2-35158-048-6); *Eds. G. De Schutter and K. Audenaert*

Report 39: Final Report of RILEM Technical Committee TC 187-SOC 'Experimental determination of the stress-crack opening curve for concrete in tension' (ISBN 978-2-35158-049-3); *Ed. J. Planas*

Report 40: State-of-the-Art Report of RILEM Technical Committee TC 189-NEC 'Non-Destructive Evaluation of the Penetrability and Thickness of the Concrete Cover' (ISBN 978-2-35158-054-7); *Eds. R. Torrent and L. Fernández Luco*

Report 41: State-of-the-Art Report of RILEM Technical Committee TC 196-ICC 'Internal Curing of Concrete' (ISBN 978-2-35158-009-7); *Eds. K. Kovler and O. M. Jensen*

Report 42: 'Acoustic Emission and Related Non-destructive Evaluation Techniques for Crack Detection and Damage Evaluation in Concrete'—Final Report of RILEM Technical Committee 212-ACD (e-ISBN: 978-2-35158-100-1); *Ed. M. Ohtsu*

Report 45: Repair Mortars for Historic Masonry—State-of-the-Art Report of RILEM Technical Committee TC 203-RHM (e-ISBN: 978-2-35158-163-6); *Eds. Paul Maurenbrecher and Caspar Groot*

Report 46: Surface delamination of concrete industrial ffioors and other dura- bility related aspects guide—Report of RILEM Technical Committee TC 268-SIF (e-ISBN: 978-2-35158-201-5); *Ed. Valerie Pollet*

Preface

This State-of-the-Art-Report (STAR) summarizes the work of RILEM TC 251-SRT "*Sulfate Resistance Testing*" between 2013 and 2019. This STAR includes three chapters associated with the sub-committees (WPs) of the TC (WP1: *Review on sulfate attack mechanisms*, WP2: *Field aspects and observed cases*, and WP3/WP4: *Review of lab test methods and recommendations for testing, including general conclusion and further steps needed*). More precisely, these chapters are dedicated to the following topics:

Chapter 2 ("Review of Sulfate Attack Mechanisms") presents a review and a brief analysis of the underlying mechanisms, not only involved in lab tests, but that also occur in real conditions. This chapter focuses on the issues that directly influence the choice of parameters and techniques for ESA testing. The specific aspect of thaumasite (e.g., favored by low temperature) has been regarded, keeping in mind that thaumasite formation induces different degradation and needs specific assessment. Possible combination with other sources of degradation (cases of seawater, including competition between chlorides and sulfates in the binding process with the cement matrix, dry climates where damage induced by salt crystallization occurs, …) has also been addressed. The complex interplay between physical and chemical processes has thus been pointed out.

Chapter 3 ("External sulfate attacks in the field") constitutes a report on the gathered field results and relevant information on structures in field conditions and trials on exposure sites. More than 50 references related to ESA in field conditions have been analyzed. Experience in different countries/continents is thus reported, pointing out their specificities with regard to the various types of sulfate attacks. On the basis of this review, an analysis has been carried out, in order to identify the most common situations (according to the environment, the type of structures, etc.). The influence of various parameters on sulfate attacks in structures in service has been studied on the basis of the cases observed. Combined effects with other attacks have again been highlighted. The specific case of soils as an external source of sulfates has been detailed. The exposure classes adopted with respect to sulfate concentration in soils and groundwater, according to various international standards, have been summarized. The methods of mitigation of the effects of ESA and acid attacks are also reported.

Chapter 4 ("Review of existing lab test methods for External Sulfate Attacks and Recommendations for an improved testing methodology") includes a critical analysis of existing lab test procedures dedicated to the assessment of ESA resistance in various countries. A focus on recent improvements is also made. In addition, promising researches dedicated to improved/innovative methods are described. Moreover, recommendations are detailed for an improved testing methodology, considering various cases (depending e.g. on the purpose), with regard to the various aspects of ESA, and including a set of test methods in order to assess various complementary parameters and not only volume change of prisms/bars, and for both mortar and concrete samples. The proposed methodology and protocols are applicable in particular to mixtures which incorporate high volumes of SCM (with Portland clinker) and more generally to various types of low-CO_2 binders. Suggestions for further steps are also provided in this chapter, such as to carry out Round Robin Tests in order to investigate deeply the role of some parameters, to compare various options in the protocols, and to assess repeatability/reproducibility data relatively to the recommended test methods. The need for relevant specific durability indicators and associated criteria is also mentioned.

This STAR will contribute to a better understanding of ESA and associated degradations, not only involved in accelerated (lab) tests, but also occurring in various complex field conditions. Moreover, this document includes valuable information and useful recommendations for the establishment of improved test procedures in the field of External Sulfate Attacks and for their future standardization.

Marne-la-Vallée, France — Véronique Baroghel-Bouny
Madrid, Spain — Esperanza Menéndez Méndez
Chennai, India — Manu Santhanam

Acknowledgments

This State-of-the-Art-Report (STAR) summarizes the work of RILEM TC 251-SRT "Sulfate Resistance Testing" between 2013 and 2019. This volume is one of the main deliverables of the RILEM Technical Committee. More precisely, it covers the various aspects developed in the working groups of this Technical Committee, such as sulfate attack mechanisms, field behavior, lab test methods, and recommendations for testing, and will thus provide an overview of the current knowledge and practice on the topic.

This book is the result of the joint effort of TC members, for drafting, reading, and reviewing. I am very grateful to them and in particular to the WP and Chapter leaders for the hard job to organize and finalize the work.

In addition, I take this opportunity to sincerely thank the local organizers of the RILEM workshops organized by the TC:

- LNEC (Lisbon, Portugal) and in particular ***Isabel Martins*** for the organization of the *Workshop on "External Sulfate Attack in concrete structures" (ESA 2016)* held on November 3–4, 2016,
- IETcc-CSIC (Madrid, Spain) and in particular ***Esperanza Menéndez Méndez*** for the organization of the Final Open Workshop "*External Sulfate Attack—Field aspects and Lab Tests*" *(TESA 2018)* held on May 24–25, 2018.

I would like to add special thanks to all the attendees of the workshops and meetings we had during the TC lifetime, for their interest and active contribution to the lively and fruitful discussions. They have thus contributed to the scientific quality of the proceedings and of the present book.

I hope that this STAR will contribute to a better understanding of ESA and associated degradations, not only in accelerated (lab) tests, but also in various complex field conditions.

Moreover, this document is likely to provide valuable information and useful recommendations for the establishment of improved test procedures in the field of External Sulfate Attacks and for their future standardization.

Véronique Baroghel-Bouny
Chair of RILEM TC 251-SRT
University Gustave Eiffel, Marne-la-Vallée, France

Contents

1 **General Introduction** 1
Véronique Baroghel-Bouny, Esperanza Menéndez Méndez, and Manu Santhanam

2 **Review of Sulfate Attack Mechanisms** 7
Manu Santhanam, Fred Aguayo, Carmen Andrade, Véronique Baroghel-Bouny, Johannes Haufe, Ashlee Hossack, Klaus Lipus, and Esperanza Menéndez Méndez

3 **External Sulfate Attacks in the Field** 49
Esperanza Menéndez Méndez, Véronique Baroghel-Bouny, Ricardo García-Roves, Isabel Martins, Florian Mittermayr, Yury Andres Villagrán-Zaccardi, and Claudio Zega

4 **Review of Existing Lab Test Methods for External Sulfate Attacks and Recommendations for an Improved Testing Methodology** 117
Véronique Baroghel-Bouny, Myriam Carcassès, Franck Cassagnabère, Nele De Belie, Chiara Ferraris, Fredrik P. Glasser, Li-ping Guo, Andreas Leemann, Klaus Lipus, Ahmed Loukili, Mathias Maes, Esperanza Menéndez Méndez, Samuel Meulenyzer, Florian Mittermayr, Othman Omikrine-Metalssi, Antonio Princigallo, Emmanuel Roziere, Manu Santhanam, Karen Scrivener, Anya Vollpracht, and Kai Tai Wan

Chapter 1
General Introduction

Véronique Baroghel-Bouny, Esperanza Menéndez Méndez, and Manu Santhanam

Abstract This chapter summarizes the work of RILEM Technical Committee 251-SRT "Sulfate Resistance Testing" between 2013 and 2019. It describes in particular the framework, objectives and structure of this technical committee, as well as the main deliverables. The various aspects developed in the working groups of this Technical Committee are indicated, such as sulfate attack mechanisms, field behavior, lab test methods, and recommendations for testing. The content of the whole State-of-the-Art-Report, which is one of the main deliverables of the TC and provides an overview of the current knowledge, standardization and practice on the topic of External sulfate Attack, is also mentioned.

Keywords RILEM TC 251-SRT · ESA · Framework · Presentation · Objectives · Structure · Deliverables

V. Baroghel-Bouny (✉)
University Gustave Eiffel, Marne-la-Vallée, France
e-mail: veronique.baroghel-bouny@univ-eiffel.fr

E. Menéndez Méndez
IETcc-CSIC, Madrid, Spain

M. Santhanam
Indian Institute of Technology Madras, Chennai, India

V. Baroghel-Bouny et al. (eds.), *External Sulfate Attacks - Mechanisms, Field Behavior and Lab Test Methods*, RILEM State-of-the-Art Reports 48,
https://doi.org/10.1007/978-3-032-19402-2_1

1.1 Framework

External sulfate attack (ESA) affects the long-term durability of various types of concrete structures.

ESA is not a new topic and, when RILEM TC 251-SRT started, substantial research work had already been done on this topic in various projects and working groups, e.g., within the framework of the European standardization (Task group CEN/TC51-TC104/WG12/TG1 "Sulfate resistance testing", led by Dr. V. Baroghel-Bouny, Univ. Gustave Eiffel, France, and formerly by Dr. G. Moir, Blue Circle, UK), the European Nanocem Consortium (led by Prof. K. Scrivener, EPFL, Switzerland) and associated MC-ITN projects, and other previous RILEM TCs (e.g., RILEM TC 211-PAE "Performance of cement-based materials in aggressive aqueous environments", chaired by Prof. M. Alexander, Univ. of Cape Town, Rondebosch, South Africa). A summary of the main inputs of these projects and working groups are included in the present STAR, in particular in the literature review of Chaps. 2 and 4.

Although ESA has been studied for decades and significant advances have been made, this topic is complex and may still be regarded as a controversial and confusing topic. Moreover, tools dedicated to accurately quantifying and predicting the phenomena and possible structural degradations are still needed.

1.2 Objectives and Challenges of the TC

RILEM TC 251-SRT was proposed under the suggestion of the CEN/TC51-TC104/WG12/TG1 "Sulfate resistance testing" with the primary aim to provide recommendations regarding appropriate test methodology and protocols dedicated to the analysis and the assessment of sulfate resistance, suitable to new types of cements, binders and concretes.

An additional objective was to provide matter for a better understanding of the underlying mechanisms, which take place, not only in concrete samples during accelerated lab tests, but also in concrete structures submitted to combined actions in field conditions. This will in particular help in the development of tools to properly relate results in lab with actual behaviour in natural environment.

Academics, testing laboratories, regulators, industrialists and practitioners will all be able to make use of the recommendations provided by this TC, more specifically the CEN community. Moreover, these recommendations will favour the safe use of "green" materials by ensuring their long-term durability to ESA.

1.3 Presentation of the TC and of Its Activities

RILEM TC 251-SRT "Sulfate Resistance Testing" deals with ESA of concrete samples, buildings and structures.

This committee is chaired by:

- Dr. V. Baroghel-Bouny (Univ. Eiffel, Marne-la-Vallée, France) *(Chair)*
- Dr. E. Menéndez Méndez (IETcc-CSIC, Madrid, Spain) *(Deputy Chair).*

It belongs to **Cluster B** "*Transport and deterioration mechanisms*" within the RILEM organization.

Members have been recruited from academic and research institutes, as well as from the industry.

According to the member affiliation during the life of the TC, RILEM TC 251-SRT regroups 32 members (see Fig. 1.1a), where most of them are from Europe (25 members) (see Fig. 1.1b). The same number of participants is from Americas and Asia (3 members). The TC includes also 1 member from Africa. Other peoples also joined the workshops and/or contributed to the drafting of this STAR.

The TC was launched in September 2013 for a 5-year period. Over the period 2013–2018, 10 meetings were held (2 per year). The kick-off meeting took place on the 3rd of September 2013 in Paris, France, and the wrap-up meeting was held on the 25th of May 2018 in Madrid, Spain. The TC was closed in 2019.

During the life of the TC, three workshops were organized:

- *Mini-Workshop on External sulfate attack*, in connection with ICCC2015 Conference (Oct. 15, 2015, Beijing, China),
- *Open Workshop on External sulfate attack in concrete structures (ESA 2016)* (Nov. 3–4, 2016, Lisbon, Portugal),

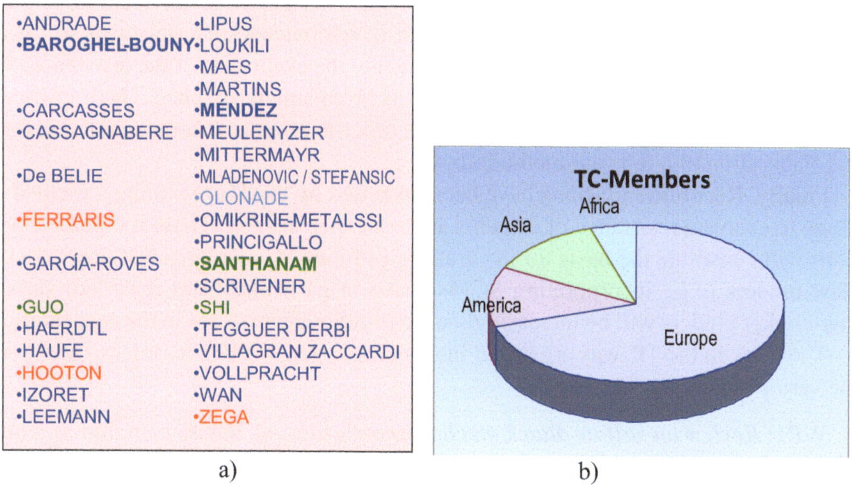

Fig. 1.1 List of members (**a**) and geographical distribution of members (**b**) of RILEM TC 251-SRT

- *Final Workshop of RILEM TC 251-SRT "External sulfate attack - Field aspects and lab tests" (TESA 2018)* (May 24–25, 2018, Madrid, Spain).

These events were organized in connection with TC meetings and provided a forum in which people from around the world came together to share their ideas and experience.

In addition, the activities and deliverables of RILEM TC 251-SRT were presented at the *Technical Day* on 29 August 2018, during the *72nd Annual RILEM Week*, which was held on 26–29 August 2018, in Delft, The Netherlands.

1.4 Scope, Contents and Structure of the TC and of the STAR

RILEM TC 251-SRT "Sulfate Resistance Testing" and the present STAR were based on the most recent advances in research related to ESA. They focus in particular on novelties in the understanding of basic **mechanisms** and of the influence of mix and environmental parameters on them, in view of improving the adequacy between ESA mechanisms/degradations and the methods of assessment of the material/structure resistance.

In addition, the TC work and the STAR include case studies of concrete structures affected by external sulfate attacks around the world. The link between **field performance** and lab tests in accelerated conditions has been addressed. The numerous data collected during the TC life from the actual cases have been analyzed as a function of the influencing parameters. The possible **coupled effects** have in particular been pointed out.

Moreover, a review and a critical analysis of the existing **lab test methods (in accelerated conditions)**, standardized or not, have been carried out. In addition, the TC work and the STAR address the recent developments undertaken in various countries with regard to test methods dedicated to the evaluation of the resistance of cements/concretes exposed to ESA, as well as promising researches. The improvements still needed have been investigated, in order to propose recommendations for the elaboration of a relevant methodology.

Finally, **Recommendations** have been proposed by the TC regarding a methodology to evaluate resistance of cements and concretes to ESA. These recommendations will constitute the basis for the drafting of future standardized test method(s). New binders (e.g., incorporating SCMs) have in particular been regarded, since low-clinker binders will be increasingly used in concrete mixtures in the near future.

The work in the TC was organized in sub-committees (WPs), in order to address the various mentioned aspects:

- **WP1:** ***Review on sulfate attack mechanisms*** (leader: M. Santhanam, India). The work issued from this WP1 constitutes Chap. 2 of the present STAR,

- **WP2:** ***Field aspects and observed cases*** (leader: E. Menéndez Méndez, Spain). The work issued from this WP2 constitutes Chap. 3 of the present STAR,
- **WP3:** ***Review of existing lab test methods for external sulfate attacks*** and **WP4:** ***Recommendations for an improved testing methodology*** (leader: V. Baroghel-Bouny, France, co-leader: A. Princigallo, Italy). The work issued from these WPs constitutes Chap. 4 of the STAR.

1.5 Deliverables of the TC

The following deliverables have been produced:

- **Proceedings** of *Open Workshop on External Sulfate Attack in concrete structures (ESA 2016)*, held on Nov. 3–4, 2016, Lisbon, Portugal (published by LNEC in 2017, Ed. by V. Baroghel-Bouny, I. Martins & E. Menéndez, and available in the Workshop website: http://esa2016.lnec.pt),
- **Proceedings** of *Final Workshop of RILEM TC 251-SRT "External Sulfate Attack -Field aspects and Lab Tests" (TESA 2018)*, held on May 24–25, 2018, Madrid, Spain (published by Springer, as: RILEM Bookseries, Vol. 21, Ed. by E. Menéndez & V. Baroghel-Bouny, RILEM, 2020, 163 p),
- Selected **papers** in Scientific Journals. These papers were derived from the contributions presented at the Workshops and have been published during the lifetime of the TC,
- the present RILEM **STAR** covering the various aspects developed in RILEM TC 251-SRT and more precisely in WP1, WP2 and WP3/WP4.

Chapter 2
Review of Sulfate Attack Mechanisms

Manu Santhanam, Fred Aguayo, Carmen Andrade, Véronique Baroghel-Bouny, Johannes Haufe, Ashlee Hossack, Klaus Lipus, and Esperanza Menéndez Méndez

Abstract The agenda of the Work Package WP1 of the RILEM Technical Committee for Sulfate Resistance Testing was to provide a review on the mechanisms of deterioration of cementitious systems in aggressive solutions bearing sulfates. It is not the purpose of this chapter to be a treatise on sulfate attack. Rather, it mainly discusses issues that directly influence the choice of testing techniques and parameters for sulfate attack. The contents are divided into three sections.

Keywords Durability of cementitious materials · Transport mechanisms · External sulfate attack mechanisms · Chemical attack · Salt crystallization

---

M. Santhanam (✉)
IIT Madras, Chennai, India
e-mail: manus@iitm.ac.in

F. Aguayo
University of Washington, Seattle, USA

C. Andrade
CIMNE, Barcelona, Spain

V. Baroghel-Bouny
University Gustave Eiffel, Marne-la-Vallée, France

J. Haufe
Zert plus (former RWTH Aachen University), Aachen, Germany

A. Hossack
University of New Brunswick, Fredericton, NB, Canada

K. Lipus
VDZ, Düsseldorf, Germany

E. Menéndez Méndez
IETcc-CSIC, Madrid, Spain

V. Baroghel-Bouny et al. (eds.), *External Sulfate Attacks - Mechanisms, Field Behavior and Lab Test Methods*, RILEM State-of-the-Art Reports 48,
https://doi.org/10.1007/978-3-032-19402-2_2

2.1 Introduction to External Sulfate Attack (ESA)

External sulfate attack (ESA) refers to the alteration of cement paste structure by sulfates penetrating the cementitious matrix from the external environment. Apart from the pore structure and chemistry of the cementitious matrix, the characteristics of the sulfate solution also significantly affect the process. In addition, the external environment—temperature, humidity, pH etc. can add to the complexity of the process.

The purpose of this chapter is to identify the specific transport mechanisms associated with external sulfate attack, and then specifically deal with the types of aggressive sulfate solutions that are typically known to deteriorate concrete. The details of the chemical reactions and physical phenomena involved in the process are provided next, followed by an understanding of the reaction products and their specific characteristics. Finally, a description of the influence of the binder type and w/b on the sulfate resistance is provided. The discussions in the various sections are focussed towards understanding the implications of different forms of sulfates and different material related parameters on the choice of the testing methods for ESA as well as the interpretation of these results.

2.1.1 Transport Mechanisms

2.1.1.1 Permeation

Permeation occurs when sulfate bearing waters penetrate concrete under pressure, and the rate/extent of penetration is determined by the permeability of the concrete. Permeation involves the entry of all types of soluble aggressive agents into concrete from the external environment.

2.1.1.2 Diffusion

Diffusion occurs as a result of the difference in ionic concentrations in the concrete and in the service environment. The diffusion of sulfates in concrete mainly follows a non-steady state process, as the interaction of sulfates with hydrated cement paste leads to the deposition of attack products such as gypsum and ettringite, which alters the pore structure of the concrete. A schematic illustration of diffusion is shown in Fig. 2.1.

2.1.1.3 Migration

Migration is the movement of ions through a porous medium due to the difference in electrical potential. In actual field conditions, it may be difficult to get such a scenario. Nevertheless, migration often forms the basis for several durability test

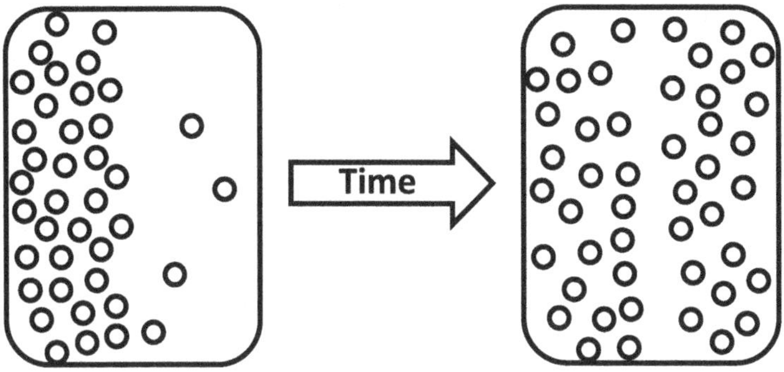

Fig. 2.1 Schematic illustration of the diffusion process

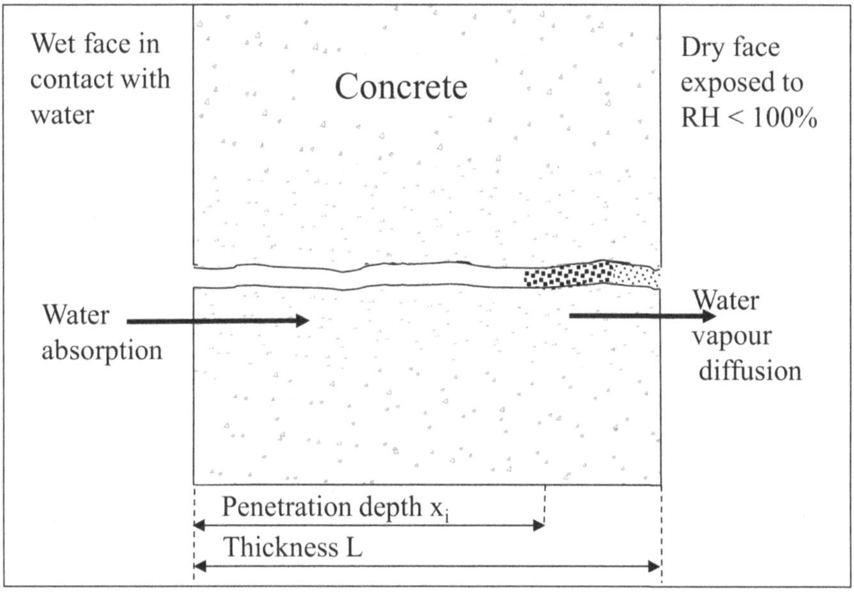

Fig. 2.2 Schematic of wick action (Buenfeld et al., 1995)

methods for concrete, the most popular among these being the Rapid Chloride Permeability Test (ASTM C1202).

2.1.1.4 Wick Action

Wick action is the transport of liquid through a porous element from a face in contact with water to a drying face (Buenfeld et al., 1995). Ions dissolved in the liquid are precipitated as salts in the pores or on the surface of the medium (Alexander et al., 2011). Wick action is a combination of other transport mechanisms such as capillary absorption, water vapour diffusion and evaporation. Figure 2.2 represents the schematic of wick action (Buenfeld et al., 1995).

2.1.1.5 Absorption

Absorption is the process by which a liquid is drawn into and tends to fill permeable pores in a porous solid body; also, it refers to the increase in mass of a porous solid body resulting from the penetration of a liquid into its permeable pores (ASTM C125, 2003). Absorption depends on pore structure as well as the moisture condition of the sample (Basheer et al., 2001). It generally occurs in unsaturated or partially saturated concrete. Different methods like cold water saturation, boiling water saturation, vacuum saturation etc. are available to measure absorption of concrete (Safiuddin & Hearn, 2005). Out of the different methods, vacuum saturation method is considered to be comparatively better. Absorption depends on total porosity of the sample.

Adsorption
Adsorption is the process in which molecules of a liquid or gas adhere to the solid. The higher energy of the solid surface leads to the attraction of foreign atoms or molecules that have a slight affinity for the surface material. If the molecules adhere to the surface only because of physical forces such as van der Waals' bonding, it is known as physical adsorption. In the case of chemisorption, the molecules react chemically with the surface, leading to much stronger bonds. In the context of concrete, at low RH, water molecules get adsorbed onto the pore walls of concrete.

2.1.1.6 Combined Mechanisms

In field conditions, there is a combination of transport mechanisms that operate together under a given condition. For example, in the case of sea water attack, for concrete exposed to the tidal zone, there is diffusion, permeation, sorption, and wick action combined with mechanical degradation brought about by the wave impact, as well as salt crystallization due to drying and wetting. Therefore, the design of test methods in the laboratory should take into cognizance this issue.

2.1.2 Aggressive Media Containing Sulfates

2.1.2.1 Gypsum and Sulfate Solutions Containing Calcium

Calcium sulfate is considered as the least aggressive of the different types of sulfates that can cause damage in the concrete (Skalny et al., 2002). This is due to the low solubility (1440 ppm) as compared to sodium and magnesium sulfates (Drimalas et al., 2011).

There are two schools of thought with respect to calcium sulfate attack:

1. The low solubility of gypsum limits the concentration of sulfate in the groundwater because of which the damage is limited (Rebel et al., 2005).

2. When the sulfates ingress into the concrete, their solubility is increased when they come into contact with the alkaline pore solution. And under these conditions, the higher concentrations of sulfates can attack more aggressively the hydrated cement paste (Hansen & Pressler, 1947).

Calcium sulfate reacts with the monosulfoaluminate and water to form ettringite that can produce expansion and cracking. It can react also with the tricalcium aluminate to form ettringite (Drimalas et al., 2011). Although most studies attribute the expansion in sulfate attack to ettringite formation, there is also evidence of expansion caused due to gypsum formation because of the primary reaction of the sulfates with calcium hydroxide and secondary reactions with C-S-H (as in magnesium sulfate attack). Tian and Cohen (2000) studied the expansion of C_3S paste caused by the formation of gypsum during sulfate attack. They observed significant expansion and loss of mass and cohesion due to the formation of gypsum.

Several works have demonstrated that the precipitation of gypsum also produces expansion (González & Irassar, 1997; Santhanam et al., 2003a, b; Tian & Cohen, 2000; Whittaker & Black, 2015). The precipitation of gypsum depends of the quantity of sulfates and the pH of the solution. The increase of pH increases the content of sulfates in the solution and favors the precipitation of gypsum.

As the sulfates penetrate into the paste, the $Ca(OH)_2$ is dissolved releasing Ca^{2+} and OH^- (Adenot & Buil, 1992; Kamali et al., 2003; Planel et al., 2006; Revertegat et al., 1992; Roziere & Loukili, 2011; Whittaker & Black, 2015).

The pH of the pore solution decreases with the dissolution of different species and results in the decalcification of the calcium-silicate-hydrate. Further, a pH value lower than 10.7 leads to the decomposition of the ettringite, transforming it to gypsum (Gabrisova et al., 1991; Whittaker & Black, 2015). In Fig. 2.3, the predicted solubility isotherms of gypsum, hemihydrate and anhydrite in 1.5 mol/kg of H_2O of $MnSO_4$ aqueous solution are presented.

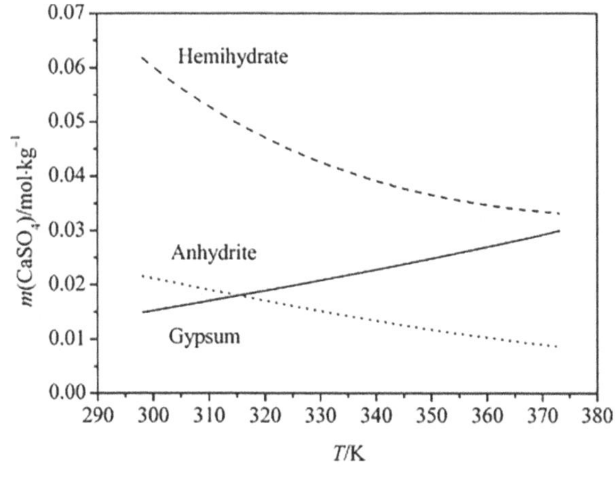

Fig. 2.3 Solubility isotherms of gypsum, hemihydrate and anhydrite (Wang et al., 2015)

2.1.2.2 Sulfate Solutions Containing Sodium

The attack of sodium sulfate is more complex than the attack of calcium sulfate, because more phases are affected by it. The attack can progress in the following ways:

1. Sodium sulfate reacts with the calcium hydroxide (portlandite) to form gypsum. The gypsum then reacts with the monosulfoaluminate to form ettringite. When the calcium hydroxide is consumed, the excess of gypsum is formed in the system and the formation of ettringite stops (Gollop & Taylor, 1995).
2. Sodium sulfate reacts with the tricalcium aluminate to form ettringite (Drimalas et al., 2011).

Yu et al. (2013) identified three different zones in the profile of sulfates, when mortars of Portland cements are submerged in sodium sulfates:

- An external surface without sulfates, due to the absence of calcium caused by leaching
- An increasing of concentration until a maximum of a depth of 0.5–1 mm
- A gradual decrease of concentration of sulfates until the background level at a depth of several millimeters.

However, in their study, the depth of penetration varied with the size of the sample and the concentration of the solution.

More detailed discussion of the phenomena involved in physical salt attack involving sodium sulfate is presented in a later section.

2.1.2.3 Sulfate Solutions Containing Magnesium

The attack of this type of sulfate is the most complex, because it can react with all the hydrated phases. It is consider the most harmful sulfate. Magnesium sulfate reacts with the calcium silicate hydrate to form gypsum and magnesium hydroxide and silica gel. This formation of magnesium hydroxide (brucite) forms a barrier that can offer protection to the concrete and reduce the permeability (Santhanam et al., 2003; Menendez et al., 2013). But the sulfate can diffuse through the layer of brucite to form ettringite in the interior, where the pH is higher. The aggressive nature of the salt is because of the low solubility of the brucite (Whittaker & Black, 2015). The formation of brucite needs a high quantity of calcium hydroxide. Because of this, when the CH is consumed, a decalcification of C-S-H is produced (Gollop & Taylor, 1995). Figure 2.4 and Table 2.1 show the stability and the thermodynamic standard properties of hydrated sulfates of magnesium respectively (Grevel & Majzlan, 2009).

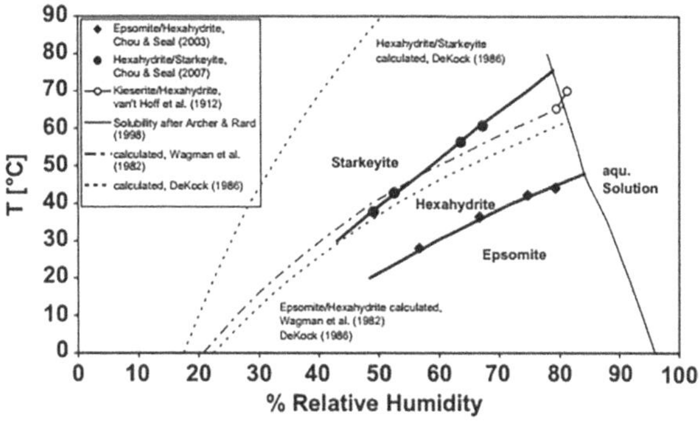

Fig. 2.4 Stability of magnesium sulfate hydrates as a function of temperature and relative humidity (Grevel & Majzlan, 2009)

Table 2.1 Thermodynamic standard properties of magnesium sulfate hydrates (Grevel & Majzlan, 2009)

Phase	Formula	$\Delta_f H^0_{298}$(kJmol^{-1})	S^0_{298} (JK^{-1} mol^{-1})	Reference
Kieserite	$MgSO_4{\bullet}1H_2O$	−1602.10	126.40	Wagman et al. (1982)
		−1610.00	126.36	DeKock (1986)
		−1609.70	124.63	Pabalan and Pitzer (1987)
		−1609.67		Ko and Daut (1979)
Sanderite	$MgSO_4{\bullet}2H_2O$	−1896.20		Wagman et al. (1982)
		−1896.61	167.36	DeKock (1986)
Starkeyite[a]	$MgSO_4{\bullet}4H_2O$	−2496.60		Wagman et al. (1982)
		−2497.01	246.86	DeKock (1986)
		−2496.65	254.76	Pabalan and Pitzer (1987)
Pentahydrite	$MgSO_4{\bullet}5H_2O$		296.83	Pabalan and Pitzer (1987)
Hexahydrite	$MgSO_4{\bullet}6H_2O$	−3087.00	348.10	Wagman et al. (1982)
		−3087.29	348.11	DeKock (1986)
		−3085.80	348.13	Pabalan and Pitzer (1987)
			348.11	Cox et al. (1955)
Epsomite	$MgSO_4{\bullet}7H_2O$	−3388.71	372.00	Wagman et al. (1982)
		−3389.04	372.38	DeKock (1986)
		−3386.94	372.41	Pabalan and Pitzer (1987)
			375.47	Gurevich et al. (2007)
Meridianiite	$MgSO_4{\bullet}11H_2O$	−4579.79		Chou and Seal (2007)

[a]Pabalan and Pitzer (1987) refer to starkeyite as leonhardtite

2.1.2.4 Sulfate Solutions with Chlorides

The sulfates found in groundwater are often a combination of calcium, sodium, and magnesium sulfate. Owing to the low solubility of calcium sulfate (~2 g/l at 20 °C), the highly soluble sulfates (Na, Mg, Fe) present in soil predominate in the groundwater solution. Because of the varying solubility of the different sulfate forms, it is important to know the amount of available sulfate in the groundwater, rather than the total amount of sulfate in the soil.

The attack of sulfates in seawater becomes complicated because of the presence of a substantial quantity of Cl ions. Chlorides penetrating into hydrated cement paste can bind the C_3A to form chloroaluminate compounds such as Friedel's salt (Verbeck, 1975). The lowered availability of C_3A can reduce the damage caused by sulfate attack due to a direct reduction in the quantity of ettringite that can form. However, no clear documentation exists of the combined effect of sulfates and chlorides. It is essential to understand the details of seawater attack, and enlist differences with groundwater sulfate attack.

Bonen and Cohen (1992a, b) investigated the effect of magnesium sulfate solutions on portland cement pastes. They suggested that the attack by magnesium ion primarily leads to the formation of a brucite layer at the surface. A layer of gypsum underneath the brucite layer is also formed, by a two-way diffusion of ions. The SO_4 ions from the solution diffuse inwards, while the OH ions from the cement paste move outwards. Because of the low solubility of brucite, the penetration of Mg^{2+} beneath the brucite layer into the interior of the paste specimen is restricted. It should be noted, however, that brucite formation consumes a high amount of CH supplied by the hydrated paste. Once the available CH is depleted, the pH of the pore solution gets lowered. In order to maintain its stability, the C-S-H can release CH to the surrounding solution, increasing the pH. This process ultimately contributes to the decalcification of the C-S-H, and the loss of the cementitious structure. In the advanced stages of attack, the Ca ion in the C-S-H can be completely replaced by the Mg ion, leading to the formation of magnesium silicate hydrate (M-S-H), which has been reported to be non-cementitious (Bonen, 1992).

When mineral admixtures such as silica fume are used, the pozzolanic reaction consumes calcium hydroxide. Thus the attack by magnesium sulfate solutions can proceed quickly to the stage where decalcification of the C-S-H gel starts occurring. This leads to a poor performance of the systems containing silica fume compared to Type I cement (Gollop & Taylor, 1995; Cohen & Bentur, 1988). However, the reduction in permeability and refinement of pore structure with the use of mineral admixtures can often overcome this negative effect. Thus, it is essential to determine critical dosage levels of the pozzolans and slag cements to maximize their benefits, and minimize the deleterious effects of magnesium sulfate attack.

2.1.3 Products of ESA and their Properties

2.1.3.1 Reactions in ESA

The sequence of reactions involved in sulfate attack has been described by Cohen and Bentur (1988). Sulfate solutions attack most of the normal hydration products of cement. Each individual hydration product can give rise to a new compound as a result of ESA.

Action on Calcium Hydroxide (CH)
Sodium sulfate ($N\bar{S}$) and magnesium sulfate ($M\bar{S}$) can react with CH to produce gypsum ($C\bar{S}H_2$), sodium hydroxide (NH) and magnesium hydroxide (MH, or brucite) according to the following approximate reactions (2.1) and (2.2):

$$CH + N\bar{S} + 2H \rightarrow C\bar{S}H_2 + NH. \quad (2.1)$$

$$CH + M\bar{S} + 2H \rightarrow C\bar{S}H_2 + MH. \quad (2.2)$$

It is not fully understood if gypsum formation causes any volumetric expansion. The formation of gypsum, however, is reported to render the structure soft (Cohen & Mather, 1991), which leads to a decrease in strength of the structure. The formation of gypsum is closely linked to the formation of other products of sulfate attack, as gypsum can combine with other hydration products to produce ettringite.

Action on Aluminates and their Hydration Products
The gypsum forming according to the reactions (2.1) and (2.2) can react further with hydrated calcium aluminates (i.e., C_4AH_{13}), hydrated calcium sulfoaluminates (i.e., monosulfate—$C_4A\bar{S}H_{12-18}$), or unhydrated tricalcium aluminate (C_3A) to produce ettringite ($C_6A\bar{S}_3H_{32}$) according to the following equations:

$$C_4AH_{13} + 3C\bar{S}H_2 + 14H \rightarrow C_6A\bar{S}_3H_{32} + CH. \quad (2.3)$$

$$C_4A\bar{S}_{12-18} + 2C\bar{S}H_2 + (10-16)H \rightarrow C_6A\bar{S}_3H_{32}. \quad (2.4)$$

$$C_3A + 3C\bar{S}H_2 + 26H \rightarrow C_6A\bar{S}_3H_{32}. \quad (2.5)$$

This phenomenon is called ettringite corrosion (Cohen & Bentur, 1988). The formation of ettringite is said to be expansive (Mehta, 1983), although the mechanism of expansion is still debated by researchers. Numerous theories exist regarding the mechanism causing expansion (Cohen, 1983; Brown & Taylor, 1999).

Action on C-S-H
According to the published literature (Cohen & Mather, 1991; Bonen & Cohen, 1992a, b), the damaging effects on the C-S-H gel are only due to the action of magnesium sulfates. The reaction proceeds as shown below in Eq. (2.6):

$$\begin{aligned} C_3S_2H_x + 3M\bar{S} + (x+21)H &\rightarrow 3C\bar{S}H_2 \\ &+ 3MH + 3SH_Y + (12 + X - Y)H. \end{aligned} \quad (2.6)$$

Thus, this reaction produces more gypsum that could go through another cycle of the reactions (2.3, 2.4, and 2.5) described earlier to produce more ettringite.

Moreover, the MH and the silica hydrate (SH_y) formed according to Eq. (2.6) can further react to produce magnesium silicate hydrate (M-S-H), which is reported to be non-cementitious (Bonen, 1992), as indicated by the reaction shown below in Eq. (2.7):

$$4MH + SH_y \rightarrow M_4SH_{8.5} + (4.5 - y)H. \tag{2.7}$$

The phenomenon of progressive reduction of the C/S ratio within the C-S-H gel is called decalcification. This process does not actually begin until the pH drops to very low values (<10).

The conversion to M-S-H is a very advanced stage of deterioration. The M-S-H and gypsum formed from the above reactions are frequently observed to be deposited in bands parallel to the exposed surface (Diamond & Lee, 1999; Bonen & Cohen, 1992a, b). The decomposition of the C-S-H gel into a non-cementitious M-S-H can be achieved only by $M\overline{S}$. In the case of $N\overline{S}$, the reaction with CH leads to the formation of NH as a product, as per Eq. (2.1). The NH thus formed compensates for any loss of alkalinity caused by the consumption of CH. However, after all the CH is consumed, the sodium sulfate solution may cause a direct attack on the C-S-H leading to the formation of thaumasite. NH is only formed up to the point of the depletion of CH. When the CH gets depleted, the decalcification of the C-S-H may eventually occur.

The decomposition by the $M\overline{S}$ doesn't stop with the formation of M-S-H but continues further (Cohen & Bentur, 1988). The action of $M\overline{S}$ renders a low pH to the pore solution. Hence the C-S-H releases some CH into the solution in order to stabilize itself at a higher pH. But since there is $M\overline{S}$ in the surrounding environment, the deterioration cycle repeats itself beginning with the gypsum corrosion.

Occurrence of thaumasite formation in sulfate attack has been reported by researchers (Erlin & Stark, 1966; Bensted, 1999). Thaumasite is believed to form during sulfate attack at low temperatures (0 to 5 °C) in the presence on CO_3^{2-} or CO_2. The crystal structure of thaumasite is still a subject of controversy, with respect to the coordination number of the Si in thaumasite. However, the general reaction for thaumasite formation can be summarized by Eq. (2.8):

$$C_3S_2H_3 + 2C\overline{S}H_2 + C\overline{C} + \overline{C} + 23H_2O \rightarrow 2C_3S\overline{C}\,\overline{S}H_{15}. \tag{2.8}$$

Thaumasite can also form as a result of the reaction of ettringite with C-S-H in the presence of CO_3^{2-} or CO_2. Thaumasite is often seen to exist in a solid solution with ettringite (Bensted, 1999; Diamond & Lee, 1999). When the attacking solution is magnesium sulfate, the formation of thaumasite can be accompanied by the formation of gypsum and brucite. This reaction (Eq. (2.9)) consumes CH in addition to C-S-H:

$$C_3S_2H_3 + 3CH + 2C\overline{C} + 4M\overline{S} + 32H \rightarrow 2C_3S\overline{C}\,\overline{S}H_{15} + 2C\overline{S}H_2 + 4MH. \tag{2.9}$$

The secondary gypsum formed in this reaction may move ahead of the brucite (Hartshorn et al., 1999), sometimes leaving gypsum and ettringite in the sound part of the specimen, while the thaumasite and brucite are discarded in the mush that spalls away because of the expansive forces generated by the chemical reactions. This could be the reason why in certain field studies, an examination of the interior part of the specimen reveals only the formation of gypsum and ettringite, leading to a false conclusion that 'conventional' sulfate attack has occurred.

2.1.3.2 Specifics of Physical Attack

According to (Lee & Kurtis, 2017; Scherer, 2004), when a porous material, such as cement or concrete, is in contact with a continuous supply of salt solution, the solution is drawn into the pores of a material by capillary suction. The flux, as a result of capillary pressure can be written in an equation derived from the Darcy's law.

$$J_c = \Phi \bar{A} \frac{dh}{dt} = \frac{k}{\eta} \frac{\left(-p_c - \rho_L g h\right)}{h} \tag{2.10}$$

where,

J_c, is the flux from the capillary rise
Φ, is the porosity of a material
$\bar{A}$, is the area of the surface in contact with the ground
k, is the permeability
η, is the viscosity,
p_c, is capillary pressure
ρ_L, is the density of the liquid
g, is the gravitational acceleration
h, is the height to which the liquid has risen

The flux decreases with the height (h) increase. On the other hand, the velocity of evaporation on the surface of a material (J_E) is a constant at any height. When the water evaporates from a porous material, the concentration of salt increases close to the material surface. Close to the area in contact with the solution, the gradient of concentration isn't big enough to drive to the super-saturation; in this region, the damage can occur when the pressure of crystallization exceeds the strength of the material.

Scherer (1999) describes the hydrostatic pressure associated with the conversion of water to ice. A large crystal of ice immersed in water at 0 °C has no tendency to grow or melt, because the chemical potentials of the ice and water are the same. But if the temperature is decreased the ice tends to grow, because its chemical potential is lower than the water. An imposed compressive stress can prevent the growing of crystals. The chemical potentials of the liquid and crystal (μ_l and μ_c) are related to

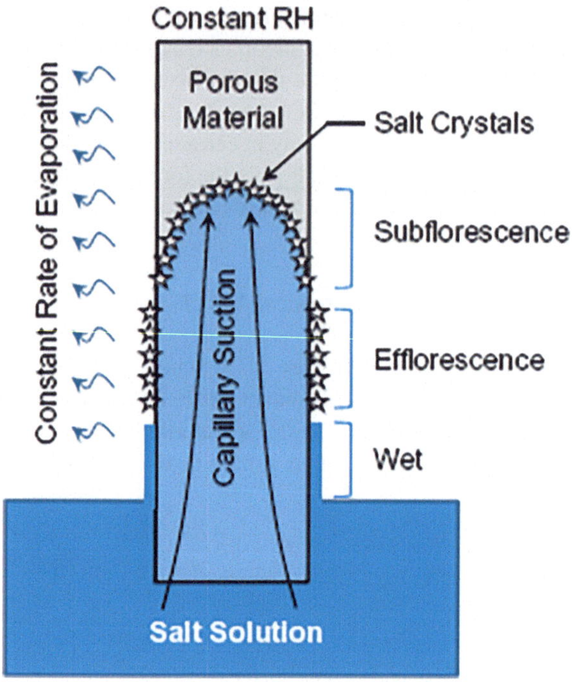

Fig. 2.5 Scheme of capillary rise and evaporation (Scherer, 1999)

their respective molar volumes (v_i and v_c) and molar entropies (S_i and S_c) by the Gibbs-Duhem equation:

$$d\mu_i = -S_i dT + v_i dp_i \tag{2.11}$$

$$d\mu_c = -S_c dT + v_c dp_c \tag{2.12}$$

At equilibrium, $d\mu_i = d\mu_c$ if pressure is applied to the crystal but not to the liquid ($dp_i = 0$), then the following equation is valid:

$$\int_{p_e}^{p_A} dp_c = \int_{T_m}^{T} \left(\frac{S_c - S_i}{v_c} \right) dT \tag{2.13}$$

where,

T_m, is the melting point of a large, unstressed crystal at pressure p_e (the pressure of the triple point, where liquid crystal and vapor are in equilibrium)
T, is the equilibrium melting temperature under applied pressure p_A

It is demostrated that the work done by the mechanical pressure is equal to the free energy difference between the crystal and liquid. Figure 2.5 shows the scheme of capillary rise and evaporation.

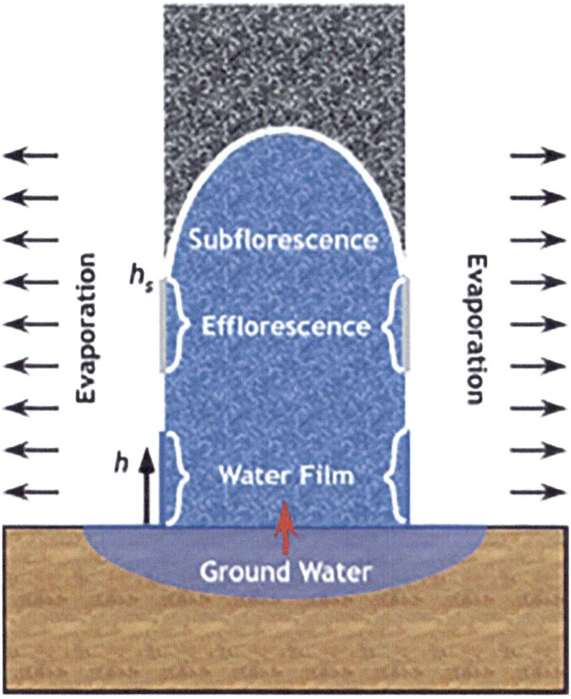

Fig. 2.6 Scheme of capillary ascent (Scherer, 1999)

The salts can be inherent to the material, be absorbed with the water or have other origins. The capillarity ascent occurs due to the capillary pressure $p_c(<0)$, that depends on the distribution of the pore sizes and the humidity inside the pore. In Fig. 2.6, a scheme of the capillary ascent is shown.

The flux into a wall is given by Darcy's law:

$$J = \Phi A \frac{dh}{dt}\frac{k}{\eta}\frac{\left(-p_c + \rho_L g h\right)}{h}\nabla p \tag{2.14}$$

where,

k, is the permeability
η, is the viscosity
p, is the density of the liquid
A, is the area of the surface in contact with the ground
Φ, is the porosity of the material
h, is the height
p_L, is the density of the liquid
g, is the acceleration of gravity

When the underground water has dissolved salts and the concentration of salt in the liquid close to the surface is increasing, the concentration of the salt on the drying surface increases continuously until the supersaturation is high enough to cause a precipitation of the salt.

According to Scherer (1999), the crystals that grow from the dissolution have a different chemical potential with the liquid:

$$\Delta\mu = R_g TLn\left(\frac{a}{a_0}\right) \tag{2.15}$$

where,

R_g, is the ideal gas constant
a, is the activity of the solution
a_0, is the activity of the solution in equilibrium with the crystal

If the activity is ignored, the expression can be rewritten as the following:

$$\Delta\mu = R_g TLn\left(\frac{C}{C_0}\right) \tag{2.16}$$

If the solution is supersaturated, the hydrostatic pressure that would have to be applied to suppress the crystal growth is:

$$p_c - p_L + \frac{R_g T}{v_c} l_n\left(\frac{C}{C_0}\right) \tag{2.17}$$

where,

v_c, is the molar volume of the crystal
p_c, is the pressure inside the glass
p_L, is the pressure of the liquid

Each face of the glass will be subjected to different pressures, since in each face the dissolution has different concentration.

$$\Delta p_c = \left(\frac{R_g T}{v_c}\right)\Delta Ln\left(\frac{C(p_c)}{C_0}\right) \tag{2.18}$$

If the pressure of each face is denominated as σ_x, σ_y, etc. then the expression becomes:

$$\Delta p_c = \sigma x - \sigma y = \left(\frac{R_g T}{v_c}\right) \Delta Ln\left(\frac{C(p_c)}{C_0}\right) \tag{2.19}$$

$$C(\sigma x) = C_0 \exp\left[\frac{\sigma x v_c}{R_g T}\right] \tag{2.20}$$

Finally, the capillary pressure in the crystal is

$$p_c - p_L = -\frac{2\gamma \cos(\theta)}{r_p} \tag{2.21}$$

2.1.3.3 Ettringite

The conversion of monosulfate to ettringite is considered as non-expansive, according to (Hime & Mather, 1999; Skalny et al., 2002). The reaction is associated with a reduction in global volume. However, the precipitation of ettringite, at the expense of the monosulfoaluminate, produces a duplication of the volume of solids as well as a precipitation of gypsum.

Different works have shown that there is no direct relationship between the quantity of ettringite and the degree of expansion. Further, studies have demonstrated that the ettringite must precipitate not only in the capillary pores but also in confined spaces to produce the damage (Lothenbach et al., 2010; Kunther et al., 2013; Menendez et al., 2019; Odler & Colan-Subauste, 1999). Also, the reduction of tricalcium aluminate decreases the appearance of ettringite (Whittaker & Black, 2015).

2.1.3.4 Gypsum

The formation of gypsum in concretes exposed to external sulfates, as described in Sect. 2.1.3.1 of this publication, is widely accepted among researchers. However, the type and magnitude of harm, if any, caused by gypsum formation is not fully understood.

Gypsum occupies a volume equal to approximately 2.25 times the volume of portlandite (Kunther et al., 2013) leading some to believe that the expansion of gypsum, upon formation, leads to sulfate attack damage. Furthermore, gypsum formation diminishes the strength of hardened cement paste by transforming calcium hydroxide, and in some cases C-S-H, into the soft product which is gypsum. However, there is considerable debate as to whether gypsum formation is expansive, causing harm, or whether gypsum forms in voids which have been caused by harmful ettringite formation; Tian and Cohen (2000) discuss this debate and conclude that gypsum formation may cause tensile stress which leads to expansion and cracking.

Kunther et al. (2013) conclude that the expansive force and subsequent increase in volume and crack formation in mortars is caused by the force exerted by forming minerals in pore solution which is supersaturated with respect to ettringite. Thus, the pressure exerted by forming ettringite crystals, not the overall increase in volume, leads to expansion and cracking. High calcium and sulfate concentrations are needed to reach a high degree of supersaturation with respect to ettringite. The concentrations of calcium and sulfate are limited by the solubility of gypsum; therefore the concentration of gypsum controls the degree of supersaturation. They conclude that ettringite formation occurs and generates expansive crystallization pressure in solutions in which ettringite and gypsum co-exist.

Schmidt et al. (2009) observed gypsum formation in cracks and veins within mortar samples which were deteriorating in the presence of external sulfates. Gypsum formation was only predicted and observed in samples which were exposed to a very high sulfate concentration (44 g Na_2SO_4/L), much higher than typical field conditions. Gypsum formation did not occur at the lower study concentration (4 g Na_2SO_4/L). They conclude that gypsum forms in cracks which have been created by ettringite expansion; gypsum does not make a significant contribution to the damage.

Thus, it is commonly known that gypsum formation is correlated to harmful external sulfate attack and deterioration of concretes; however, the role of gypsum in this process is not well understood.

2.1.3.5 Thaumasite

According to Köhler et al. (2006) and Whittaker and Black (2015), the thaumasite is produced through the heterogeneous nucleation over the surface of the ettringite due to the structural similitudes of these minerals. This reaction is followed by a higher epitaxial growth of the thaumasite from the components present in the solution (Whittaker & Black, 2015). Figure 2.7 depicts the kinetics of leaching of Ca^{2+} and OH^- and the ingress of SO_4^{2-}. It can be seen that it is necessary to have a certain quantity of ettringite to quickly form thaumasite.

On the other hand, Nobst and Stark (2003) showed that the thaumasite formed is directly proportional to the quantity of Al_2O_3. Also, it is reported that there is no transformation of ettringite to woodfordite, nor from woodfordite to thaumasite; in other words, thaumasite isn't formed throughout the route of the woodfordite.

2.1.3.6 Thenardite and Mirabilite

There are two crystalline salts of sodium sulfate, thenardite (Na_2SO_4) and mirabilite ($Na_2SO_4.10H_2O$). The solution of thenardite is supersaturated with respect to the mirabilite under a temperature of 32 °C and 75% relative humidity. The

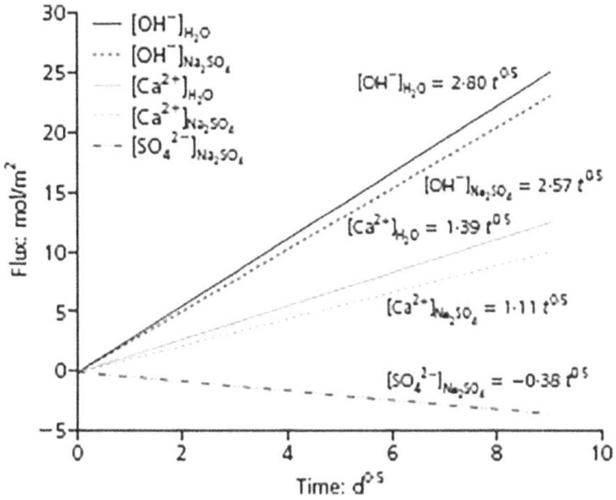

Fig. 2.7 Kinetics of leaching of Ca^{2+} and OH^- and the ingress of $SO_4^=$ (Whittaker & Black, 2015)

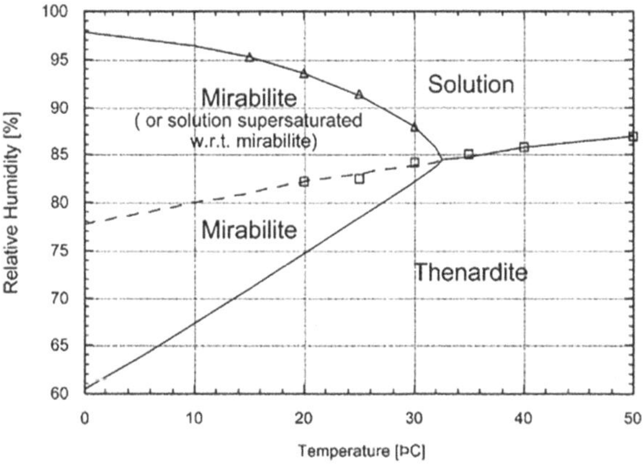

Fig. 2.8 Phase diagram of sodium sulfate—continuous lines are the boundaries of the stable phases and discontinuous line corresponds to a solution in metastable equilibrium with respect to thenardite and supersaturated with respect to mirabilite (Tsui et al., 2003)

precipitation of the mirabilite can cause damage. (Flatt, 2002; McMahan et al., 1992; Steiger & Asmussen, 2008; Rodriguez-Navarro et al., 2000) have observed that with relative humidity higher than 40% the mirabilite is formed first from a super-saturated sodium sulfate solution, and later this is converted to thenardite. At relative humidity under 40% the thenardite is directly precipitated (Thaulow & Sahu, 2004). Other authors (Correns, 1949; Scherer, 2004) have suggested as damage mechanism the crystallization of salts. This is also explained by Flatt (2002) and Whittaker and Black (2015) based on the wet and dry cycles, because during the wet phase the thenardite is solubilized and the mirabilite is supersaturated. The phase diagram of sodium sulfate, according to Tsui et al. (2003) is shown in Fig. 2.8.

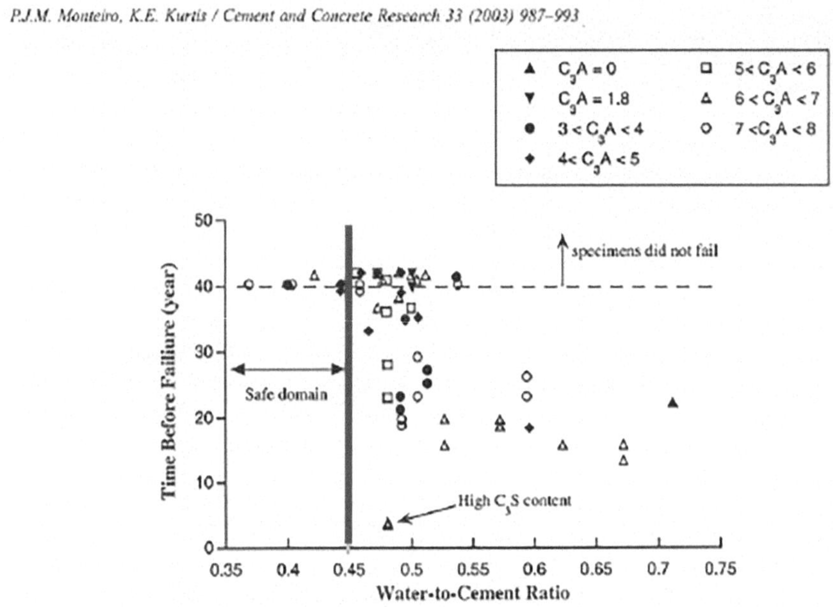

Fig. 2.9 Time to failure in exposure to sulfate solution as a function of w/c ratio and varying C_3A content of the cement (Monteiro & Kurtis, 2003)

2.1.4 Influence of Binder Type, Water-to-Binder Ratio (w/b) and Aggregates on ESA

2.1.4.1 Role of w/b

The use of a low w/c ratio results in a better resistance to external sulfate attack (Al-Akhras, 2006; El-Hachem et al., 2012; Monteiro & Kurtis, 2003). If there is a high w/c ratio the porosity and the permeability is increase, favoring the entrance of sulfates (Khatri et al., 1997). Most international codes suggest a w/c ratio lower than 0.45 to improve the resistance to the sulfates.

The 40 year non-accelerated long-term test by U.S. Bureau of Reclamation (USBR), on sulfate attack found that time to failure for concrete prepared from various mixtures, as measured by expansion, decreased with increasing w/c ratio. Monteiro and Kurtis (2003) suggested the use of concrete with w/c ratio less than 0.45 to resist sulfate attack. The time to failure due to ESA as a function of w/c ratio is shown in Fig. 2.9, and the data shows that the expected time to failure decreases with w/c ratio and increasing C_3A content. The same study also indicated that cements high in C_3S led to poor performance, presumably due to greater formation of gypsum.

The high porosity of the concrete resulted from casting with high w/c ratio may account for the resulting expansion. In some lab studies, however, the early age results may seem to favour systems with high w/c, because of a larger space available for the deposition of the attack products. In field conditions, the permeability effect dominates, and therefore the rate of attack increases with the w/c of the concrete.

The porosity of the concrete is crucial for the ability to withstand the sulfate attack. It can be stated that the porosity is likely to have a greater influence on the sulfate resistance than the C_3A content of the cement used.

The curing of the specimens is critical because external sulfate attack depends primarily on the diffusion of sulfates in the porous and permeable concrete matrix. According to (Mangat & Khaitb, 1992; Fernandez-Altable, 2009; Whittaker & Black, 2015) the specimens cured in air perform better than those cured in water because of the surface protection accorded by the carbonation.

2.1.4.2 Influence of SCMs

In general, the use of SCMs is considered beneficial to prevent the sulfate attack. Pozzolans and slag decrease the content of portlandite, albeit to different extents. The lowered availability of portlandite leads to the formation of less gypsum, and consequently less ettringite (Whittaker & Black, 2015). A content of 20% of fly ash was reported to increase the sulfate resistance (Irassar & Batic, 1989; Mangat & Khatib, 1996; Torill & Kawamura, 1994; Torill et al., 1995) due to the consumption of the portlandite (Whittaker & Black, 2015). The highly reactive pozzolans such as silica fume have been reported to perform poorly in magnesium sulfate attack (Al-Amoudi, 2002; Bonen & Cohen, 1992a, b; Lee et al., 2005).

According to ACI 201.2R-01 (Guide for Durability of Concrete), when the content of C_3A is equal to or higher than 7% as per Bogue calculations, different quantities of the fly ash, natural pozzolans, silica fume and slags are needed to increase the sulfate resistance:

1. Fly ash and natural pozzolans: replacement of 25–35% by mass
2. Silica fume: replacement of 7–15% by mass
3. Slag: replacement of 40–70% by mass

Fly Ash

Tikalsky and Carrasquillo (1992) evaluated the influence of several different fly ashes on sulfate resistance, and found that even for low C_3A content cements, the incorporation of Class C fly ash lowered the performance. It was shown that sulfate resisting Portland cement (SRPC) was better than blends of OPC with SCMs in the case of magnesium sulfate attack. The authors also stated that accelerated comparisons of performance of blended systems with OPC from mortar tests can be misleading.

By replacing cement with fly ash, the sulfate resistance of concrete is usually increased (Schubert & Lühr, 1979). This is due to the reduction in C_3A present in the binder, which is primarily responsible for the formation of expansive phases such as ettringite and gypsum (Wischers & Sprung, 1990). However, the reduction of the C_3A content is not enough to completely rule out the formation of these phases.

Schießl and Härdtl conducted investigations on the sulfate resistance of cement-fly ash combinations with a fly ash content of 20% and 40% by mass of the binder according to the SVA method (Schiebl & Härdtl, 1992). The higher porosity of the mortar due to the replacement of the fly ash did not lead to higher expansion, so it was concluded that the improving effect of the fly ash must be due to chemical-mineralogical reasons. The pozzolanic reaction of fly ash is associated with a consumption of $Ca(OH)_2$ dissolved in the pore solution. The reduction of $Ca(OH)_2$ and the correspondingly lower content of Ca^{2+} ions available for reaction were reported to reduce the expansion (Schubert & Lühr, 1979; Torill et al., 1987; Sahmaran et al., 2007).

In general, the addition of fly ash leads to a reduced expansion due to the storage in sulfate solution. With increasing storage time, the difference of expansion compared to pure Portland cement increases. Also, blast furnace cements and sulfate resistant cements showed lower expansion at a higher age when used in combination with fly ash.

Prolonged storage time resulted in lower expansion for fly-ash-containing mortars (Schubert, 1974; Schubert & Lühr, 1979). This is attributed to the densification of the pore structure by the pozzolanic reaction and the reduction in the $Ca(OH)_2$ content. However, an extension of the curing duration beyond 28 days had no measurable influence on the sulfate resistance. This contradicts the results of various authors, who were able to significantly increase the sulfate resistance by increasing the curing duration to more than 28 days (Brameshuber et al., 2000; Mulenga, 2002; Schubert et al., 1976).

The low expansion of blast furnace-cement mixtures cannot always be achieved with fly ash. This is the conclusion reached by Lang, who carried out tests using the flat prism method on 30,000 and 5760 mg/l sulfate on commercial cements of various plants and laboratory cements (Lang, 1993). Mulenga also showed that by adding fly ash of 30% by mass to Portland cement, a high sulfate resistance can be achieved, but even with an increase in fly ash content to 50% by mass of the binder, the sulfate resistance was lower than that of sulfate resistant Portland cement (Mulenga, 2002).

The sulfate resistance of sulfate resistant Portland cements can be further increased by the addition of fly ash (Schubert et al., 1976). The higher the original C_3A content of the cement is, the greater the improvement in sulfate resistance is (Schubert & Lühr, 1979). For cements with a low C_3A content, the impact of fly ash is comparatively small. If a high C_3A cement is used in combination with fly ash, the pre-storage of the concrete has a major influence on the sulfate resistance (Schubert & Lühr, 1979).

Sumer found a decrease in sulfate resistance with increasing binder content and a constant amount of fly ash in the binder when tested according to ASTM C 1012 in magnesium sulfate solution (120,000 mg/l sulfate) (Sumer, 2012).

An exchange of 25 to 41.7% by mass cement by fly ash led to an increased sulfate resistance of the mortar when stored in sodium sulfate solution of 14,200 mg/l sulfate, regardless of the C_3A content (Dikeou, 1970; Monteiro & Kurtis, 2003). The expansion increased in the following order: cement type V (C_3A content <5% by mass) and fly ash < cement type II (C_3A content <8% by mass) and fly ash < cement type V < cement type II < cement type I and fly ash < cement type I.

The increased sulfate resistance with increasing fly ash content in the range up to 50% fly ash content in the binder is general consensus in the literature (Brameshuber et al., 1999; Brameshuber et al., 2000; Ludwig et al., 2012; Mulenga, 2002; Muller-Rochholz, 1987; Ramyar & Inan, 2007; Schubert, 1974; Schubert et al., 1976; Torill et al., 1987). With a suitable combination of cement and fly ash, expansion in the range of sulfate resistant Portland cements (Brameshuber et al., 2000) and blastfurnace cements as well as combinations of cement and silica fume (Cao et al., 1997) can be achieved. Some hard coal fly ash concretes (25–33% by mass) have a much higher sulfate resistance than Portland cement concrete and concretes with sulfate resistant Portland cement (Burton & Howes, 1998), not only in sodium but also in magnesium sulfate solution containing 12,000 mg/l sulfate.

Matthews showed that after 5 years' storage in Na_2SO_4 solution and lower concentration $MgSO_4$ solution (4200 mg/l sulfate), approximately the same residual strength as concrete with SR cement can be achieved (Matthews, 1991). In (Al-Dulaijan et al., 2003) it is reported that normal Portland cement (type I) in combination with fly ash is less damaged than sulfate-resistant cement of type V even at very high sulfate concentrations up to 40,000 mg/l (Na_2SO_4). At higher concentrations, magnesium sulfate solution is more aggressive than Sodium sulfate solution (Matthews, 1991; Schubert et al., 1976).

A lower sulfate concentration in the storage solution leads to lower expansion (Ramyar & Inan, 2007), which is why many studies have been carried out at high sulfate concentrations to accelerate the sulfate attack.

In the United States, investigations were conducted on Type I, II and V cements without and in combination with fly ash (15–50% by mass) (Verbeck, 1967; Dunstan, 1980; Dunstan, 1987; Torill et al., 1987; Tikalsky & Carrasquillo, 1992). The concretes were stored in a sodium sulfate solution containing 68,000 mg/l sulfate. (Verbeck, 1967) observed no substantial improvement against sulfate attack by adding fly ash for cement types II and V, whereas for cement type I the addition of unspecified fly ash at low cement content (4 bags per cu yd $\approx$ 223 kg/m^3) resulted in less damage. With cement-rich concrete, the behavior worsened.

Fly-ash with a high content of CaO (Class C) had a negative effect on the sulfate resistance, a low content of CaO (Class F) led to an improvement (Tikalsky & Carrasquillo, 1992). In particular, high fly ash additives of up to 50% by weight of the binder lead to improved resistance, and the amount of ettringite and gypsum formed is significantly reduced. This is attributed to the reduced content of $Ca(OH)_2$, as well as the influence of the pozzolanic reaction on the microstructure (Torill et al., 1987). The pozzolanic reaction of the fly ash produces additional C-S-H phases, thus reducing capillary porosity. As a result, the tightness and thus the diffusion resistance of the structure are increased. A reduction of the water content in the concrete additionally leads to an increase of the diffusion resistance (Mulenga, 2002). An improvement of the sulfate resistance with a reduced water/cement ratio was also shown by Messad et al. for an addition of 30% fly ash after storage in Na_2SO_4 solution with 8900 mg/l sulfate (Messad et al., 2010).

The influence of the chemical composition of the fly ash was investigated by Dunstan (Dunstan, 1980). He assumed that fly ash with a chemical composition in the range of mullite in the system Al_2O_3-SiO_2-CaO does not react with sulfate. It has been postulated that a content of more the 5% in CaO worsens the sulfate resistance. Tests on a Type II cement with 15% and 25% fly ash in Na_2SO_4 solution with 68,000 mg/l sulfate were carried out to substantiate the thesis. For the evaluation of the fly ash an R-factor was introduced, which characterizes the composition of the fly ash:

$$R = (C-5)/F$$

C: CaO content of the fly ash,
F: Fe_2O_3 content of the fly ash

A good correlation of the R-factor with the measured expansion could be demonstrated. Based on the results, a classification of the fly ashes for ranges of the R-factor was proposed:

$R < 0.75$: greatly improved sulfate resistance,
$0.75 < R < 1.5$: moderately improved sulfate resistance,
$1.5 < R < 3$: no significant change,
$3 < R$: reduced sulfate resistance.

Even after many years of storage of the samples, the good correlation of the sulfate resistance with the R-factor could be confirmed (Dunstan, 1987). However, the evaluation of fly ashes by using the R-factor has not been established.

Dhole et al. showed that CaO-rich fly ashes can be used to improve sulfate resistance when combined with low-CaO fly ashes (Dhole et al., 2011) Low-CaO fly ash compensated for the negative effect of CaO-rich fly ash. Also quaternary binder combinations of cement, class F and class C fly ash and silica fume are suitable to obtain a particularly high resistance to sulfate attack. The addition of ultrafine fly ash resulted in a high sulfate resistance.

The influence of fineness of cement on the improvement of sulfate resistance is also discussed by Fraay et al. (1987). Investigations on mortar prisms of $40 \cdot 40 \cdot 160$ mm^3 with 20% Class F fly ash and storage in Na_2SO_4 solution with 2000, 13,500 and 27,000 mg/l showed that the porosity has a much greater influence on the sulfate resistance than the C_3A content and fineness of the cement.

Further investigations showed that not only the CaO content, but above all the mineralogical phase balance has an influence on the effect of the fly ash with respect to sulfate resistance (Dhole et al., 2013). In CaO-rich fly ash, reactive phases such as C_3A, gehlenite and anhydrite were identified as limiting the sulfate resistance in addition to the glass phase. By contrast, class F fly ashes consist of inert crystalline phases and aluminosilicate glass.

Table 2.2 Chemical composition of the granulated blast-furnace slags (GBFS)

	GBFS A	GBFS B	GBFS C	GBFS D	GBFS E	GBFS F
LOI	0.94	0.4	1.31	0.36	0.64	1.51
SiO_2	30.25	36.02	34.69	36.10	38.91	35.12
Al_2O_3	15.96	11.36	11.70	12.97	9.54	11.94
TiO_2	1.13	0.47	0.89	0.63	0.96	0.73
P_2O_5	0.07	0.01	0.01	0.01	0.06	0.01
Fe_2O_3	0.66	0.35	0.42	0.25	0.54	0.91
Mn_2O_3	0.39	0.48	0.33	0.31	1.18	0.34
MgO	10.00	9.36	6.80	10.30	8.45	6.98
CaO	38.68	40.28	42.30	37.54	37.57	41.28
SO_3	0.65	0.15	0.25	0.02	0.04	0.01
K_2O	0.39	0.38	0.47	0.71	1.24	0.38
Na_2O	0.39	0.27	0.27	0.26	0.36	0.27
Sulfide	0.96	0.94	1.18	1.07	1.00	1.02

All data in % by mass

In conclusion, fly ash addition to low and high C_3A Portland cement can increase sulfate resistance. However, a very wide range of fly ashes is available internationally, so it is important to distinguish between low-calcium (iron-rich) and calcium-rich fly ashes. Calcium-rich fly ash (class C) significantly worsens the resistance to sulfate attack. However, this negative effect can be counteracted by the addition of iron-rich fly ash and silica fume.

Ground Granulated Blast Furnace Slag

Cements containing more than 65% by mass of blast-furnace slag are considered in many countries to be cements with high sulfate resistance in accordance with a standard (e.g., EN 197-1 in Europe). In some national standards also cements with a lower content of granulated blast-furnace slag are considered to be high or at least moderate sulfate resistant. These cements are in Europe mainly CEM III/A, and in a few cases CEM II/B-S. The mechanism leading to the increased sulfate resistance of a concrete or mortar based on blast-furnace cement has different representations in the literature. A low C_3A content of the clinker and a low aluminate content of the blast-furnace slag are essential.

In doing so, both chemical and physical aspects are primarily held responsible. In general, however, increasing the physical diffusion resistance of the hardened cement is considered to be the main cause of the higher sulfate resistance of granulated blast-furnace slag containing cements (Al-Dulaijan et al., 2003; Brameshuber et al., 1999, 2000; Burton & Howes, 1998; Cao et al., 1997).

Research done at the VDZ showed the impact of the aluminate content of the granulated blast-furnace slag (Lipus & Baetzner, 2009). Cements containing 70%, 50% and 30% by weight blast-furnace slag were tested using the German SVA sulfate resistance test method. Here five granulated blast-furnace slags were used, which differed clearly in the chemical composition (Table 2.2).

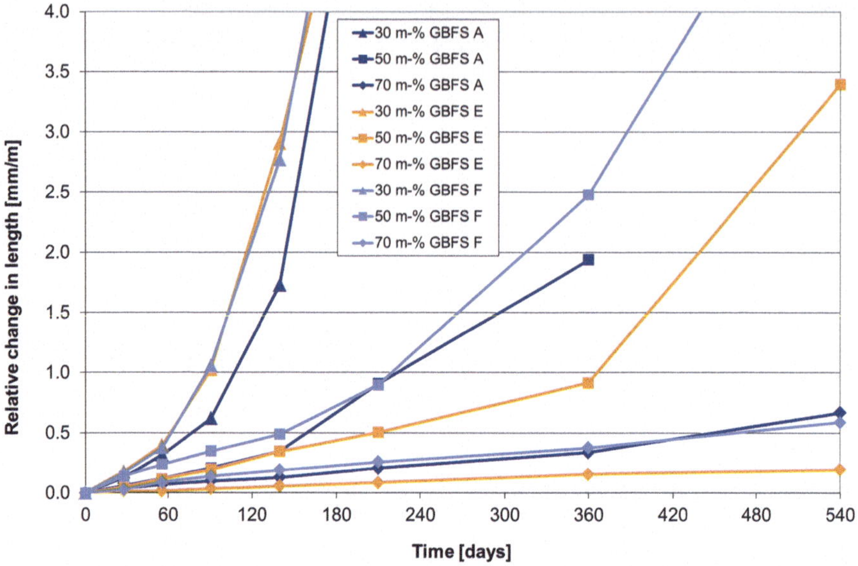

Fig. 2.10 Relative change in length of mortar specimens due to the SVA sulfate resistance test method

According to the SVA method, for cement with high sulfate resistance, the expansion of the flat prisms does not exceed the value of 0.5 mm/m within 90 days and 0.8 mm/m within 180 days. As expected, this criterion was fulfilled by the cements with 70% by mass blast-furnace slag. Likewise, the expansions of the cements containing 50% by mass of blast-furnace sludge remained below the limit. Only the cements with 30% by mass blast-furnace slag content missed the criterion as expected. With an extended test period, the number of cements increased whose test pieces showed higher expansions than 0.8 mm/m. After 360 days, alone the cements containing 70% by mass of blast-furnace slag remained below this limit value (Fig. 2.10).

According to the expansion results and the practical experience, all cements with a GBFS content of 70% by mass (CEM III/B) showed high sulfate resistance, without any influence of the slag composition. In contrast, for cements containing 50% by mass of GBFS (CEM III/A) and 30% by mass GBFS (CEM II/B-S) the chemical composition of the blast furnace slag affected the sulfate resistance in some cases very clearly (Fig. 2.11).

The sulfate resistance appears to be significantly dependent on the "silicon excess" (in relation to aluminum and calcium) of the blast-furnace slag. The larger this is, the higher is the sulfate resistance of the mortar samples produced therefrom (Fig. 2.12). Consequently, cements with low-aluminum and silicon-rich granulated blast-furnace slags are best suited for the production of SR-grade cements. For the development of strength, on the other hand, the inverse relationship applies

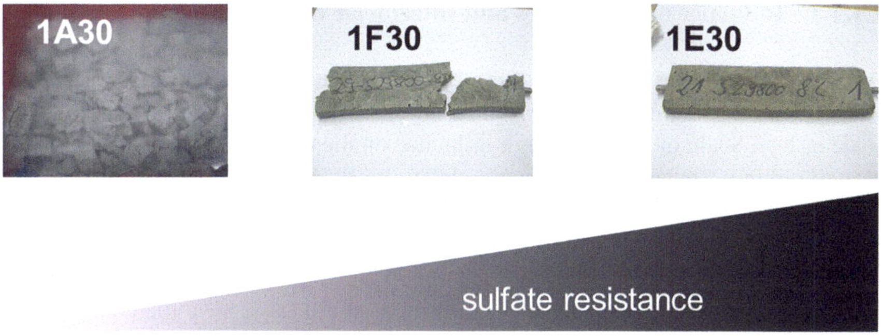

Fig. 2.11 Mortar prisms after 540 days sulfate exposure, cements with 30% by mass of GBFS

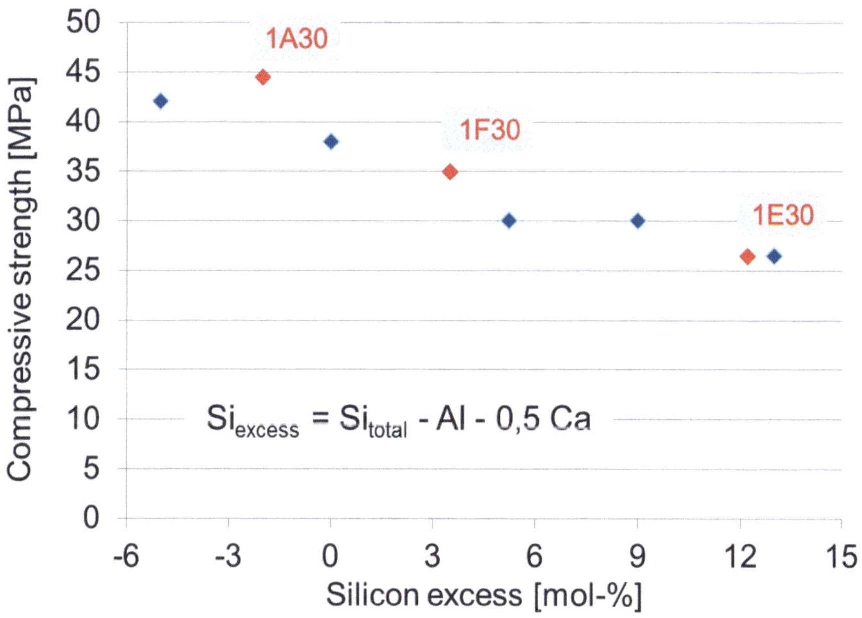

Fig. 2.12 Relationship between silicon excess and standard compressive strength (Mortar with 70% by mass GBFS)

(Wassing, 2003). SR cements with granulated blast-furnace slag contents of around 50% by mass are possible, provided that the blast-furnace slag has sufficient silicon excess (Fig. 2.12).

For the higher durability, an altered composition of the C-S-H phases seems to be responsible. Due to the available "excess" silicon the blast-furnace slag can react with calcium hydroxide to form further or altered C-S-H (lower C/S ratio). The uptake of sulfate ions is not avoided, but the forming phases do not lead to cracking and disintegration of the microstructure.

2.1.4.3 Influence of Aggregates and Interfacial Transition Zone

Another material parameter that can alter the kinetics of the sulfate attack is the aggregate size. Size of aggregate defines the Interfacial Transition Zone (ITZ) and scale of test. Scale of test has a high influence on the time span of testing. Research carried out in paste and mortar cannot always be extrapolated to get results on concrete. A difference in microstructure of the paste and the cement paste within concrete is due to the ITZ effect. Overlap of the effects of transition zones within the concrete, alters the performance of the paste in concrete compared to the individual effect of paste. Studies are typically conducted on mortar to save material and cost and to represent concrete with the presence of ITZ.

2.2 Physical, Chemical and Microstructural Alterations in Esa

2.2.1 Effects of Gypsum and Ettringite Formation

Ettringite and gypsum are the primary products of the chemical reaction between sulfate-bearing solutions and cement hydration products. This fact has been recognized for a long time, although considerable controversy exists in identifying the significance of each of these products in the ultimate deterioration of concrete undergoing sulfate attack.

Failure by expansion as a result of sulfate attack has traditionally been attributed to the formation of ettringite (Mehta, 1983; Cohen, 1983). Some researchers have not been able to detect a strong correlation between expansion and ettringite formation (Odler & Gasser, 1988). However, the potential for expansive ettringite formation depends on the type of C_3A present in the cement. C_3A is known to exist on at least four different crystalline forms in cement, which differ in reaction rates but not in the total amount of ettringite formed (Mehta, 1978). The ettringite formed with the cubic form of C_3A is found to be more expansive than the tetragonal, monoclinic, or the orthorhombic forms. It should also be noted that the expansion is primarily associated with the formation of ettringite within the bulk cement paste, and not inside voids or ITZ.

The formation of ettringite according to the reactions in Sect. 2.1.3 primarily leads to an increase in the solid volume (Glasser, 1999). The expansion due to ettringite is attributed to various theories, some of which are conflicting (Brown & Taylor, 1999). One of the main theories proposed is that the expansion is due to the crystal growth pressure during the formation of ettringite (Ping & Beaudoin, 1992; Flatt & Scherer, 2008; Kunther et al., 2015). The other school of thought is based on swelling of ettringite by intake of water in its crystal structure (Cohen, 1983).

Gypsum is the primary reaction product of sulfate attack at high sulfate-ion concentrations (>8000 ppm SO_4^{2-}) (Biczok, 1967). Ettringite is not stable in the low-lime environments present at high sulfate concentrations, when the pH falls below 11.5–12 (Biczok, 1967). At this low pH range, ettringite could decompose to form gypsum.

Even though the formation of gypsum is important, there is no clear documentation of the nature of disruption caused by gypsum (Hansen, 1963; Cohen & Mather, 1991; Tian & Cohen, 2000). The formation of gypsum according to the reactions in Sect. 2.1.3 is accompanied by a decrease in the total volume, when the volume of the products and reactants is compared. However, the solid volume actually increases as a result of the reaction. Moreover, much like ettringite, gypsum conventionally deposits in a needle-like form, with numerous air voids. A micrograph depicting gypsum formation as a result of hydration of hemihydrate ($CaSO_4.0.5H_2O$) has been shown in Fig. 2.13a. From the picture, the crystallization of gypsum in a needle-like form is evident. The original hemihydrate is actually present as large lath-like crystals. Thus a lot of voids are incorporated leading to an increase in the volume.

The study of gypsum formation is often not emphasized enough in traditional sulfate attack research. This explains the choice of low C_3A cements universally as a protection against sulfate-bearing solutions. Observations from field structures indicate that even sulfate resisting Portland cements show signs of damage in sulfate attack situations. There are two aspects of gypsum related deterioration that need to be addressed: (a) surface softening, and (b) expansion. Softening of the concrete surface and mass loss due to spalling are commonly observed in sulfate-related damage in field structures, but these effects are not usually quantified (Cohen & Mather, 1991). Standards need to be developed for quantification of the surface-softening phenomenon including mass loss and depth of damage. A hardness method could be used to detect areas of damage.

(a) Gypsum crystallizing in the needle form (b) Gypsum crystallizing in a stubby form in the presence of sodium citrate

Fig. 2.13 Gypsum forming from the hydration of hemihydrate (note: this is a hemihydrate and water system; no cement) (Santhanam, 2001). (**a**) Gypsum crystallizing in the needle form (**b**) Gypsum crystallizing in a stubby form in the presence of sodium citrate

In order to study whether gypsum formation causes expansion, it is important to use a C_3A-free system, to avoid the effects of ettringite. In a study of the properties of alite (C_3S) cements by Mehta et al. (1979), the performance of concrete specimens prepared with alite cements under sulfate exposure was investigated. Long-term exposure caused significant spalling and loss of strength in alite concrete as compared to concrete prisms prepared with an ASTM Type V cement. This phenomenon was attributed to gypsum formation, which was detected by X-ray diffraction studies. However, expansion measurements were not made in this study.

In a study by Tian and Cohen (2000), the effect of gypsum formation during sulfate attack on the resulting expansion of C_3S paste and mortar bars was investigated. The quantities of gypsum formed in various mixtures were compared by the relative heights of XRD peaks. It was found that C_3S mortars were highly susceptible to expansion, and registered a high degree of gypsum formation. This effect was reduced when the C_3S was partially substituted with silica fume.

Using thermal analysis techniques, the amounts of various phases as a mass percentage of hydrated cement paste can be determined. These techniques can be used to monitor the amounts of gypsum and ettringite formed during the reaction. Quantifying the role of gypsum and ettringite can help in identifying the optimum cost-effective solutions to mitigate sulfate-related damage. For example, crystal growth inhibitors, such as sodium citrate (which inhibits the crystal growth of gypsum) can be used as a possible additive to reduce gypsum formation (Lea & Nurse, 1949). In the presence of sodium citrate, gypsum crystallizes in a short and stubby form rather than in the needle form, as shown in Fig. 2.13b.

2.2.2 *Thaumasite Form of Sulfate Attack*

Thaumasite sulfate attack occurs due to the formation of thaumasite ($CaSiO_3.CaCO_3.CaSO_4.15H_2O$) from calcium silicate-hydrates, sulfates, and carbonates (Schmidt et al., 2008). Calcium silicate-hydrates (C-S-H) are derived from the hardened cement paste, sulfates typically come from an external sulfate source, and the carbonates can be derived from internal or external sources. Thaumasite sulfate attack destroys the C-S-H matrix of the cement paste, thereby leading to a loss in integrity of the concrete. See Fig. 2.14a, b for examples of classical and thaumasite sulfate attack, respectively.

The generally-accepted process of thaumasite formation is: sulfate ions enter the cement matrix, ettringite and gypsum are formed (classical external sulfate attack, see previous section), decalcification of C-S-H occurs, and then thaumasite is formed. Thaumasite formation is the final stage of external sulfate attack. The progress of deterioration due to thaumasite formation is discussed in greater detail in the following section.

Thaumasite sulfate attack has been observed more frequently colder temperatures (0–5 °C) (Bensted, 1999); however, thaumasite has been observed at higher

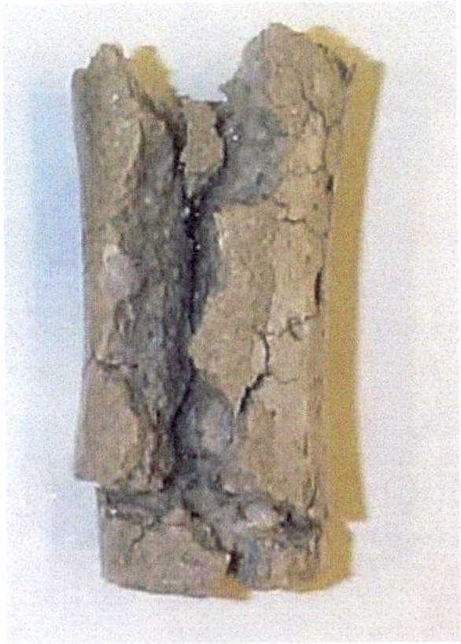

(a) Conventional sulfate attack

(b) Thaumasite form of sulfate attack [Barker and Hobbs, 1999]

Fig. 2.14 Visual appearance of failed mortar specimens. (**a**) Conventional sulfate attack. (**b**) Thaumasite form of sulfate attack (Barker & Hobbs, 1999)

temperatures (20 °C) also (Irassar, 2009). Thaumasite sulfate attack has been modelled by numerous researchers (Lothenbach et al., 2010; Schmidt et al., 2009). Thaumasite formation is a kinetically slow process. Many of the presently-available sulfate attack models assume equilibrium has been reached; they do not relate to time or distance. Sulfate attack models allow for a calculation of the potential volume of thaumasite formed although this total volume might not be formed in laboratory samples. See Fig. 2.15 for examples of the modelled volumes of minerals in the hardened cement paste as a function of the volume of sodium sulfate added to the cement paste. Note: the model assumes equilibrium, the sulfate concentrations at the surface are 4 g/L Na_2SO_4 (a), and 44 g/L Na_2SO_4 (b), and the cement contains 5% interground limestone.

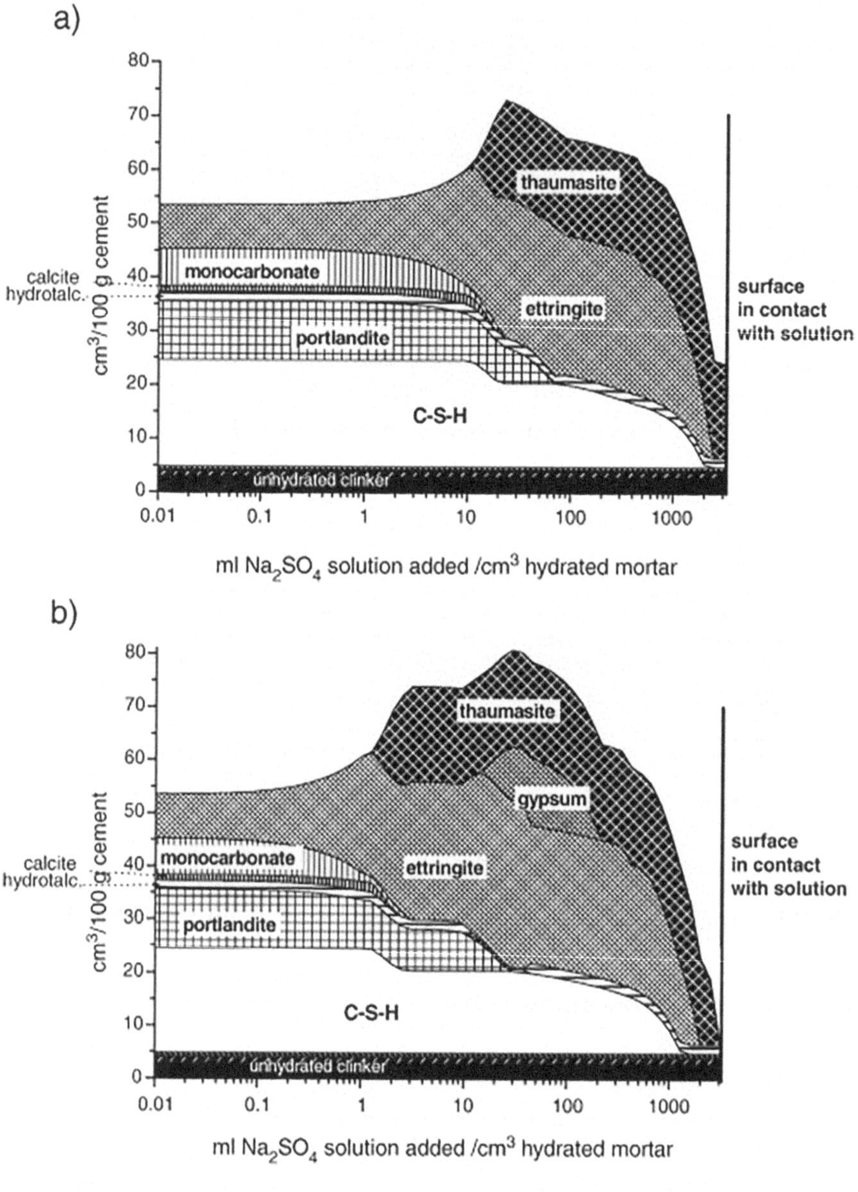

Fig. 2.15 Calculated phase assemblage of mortar sample immersed in (**a**) 4 and (**b**) 44 g/L Na_2SO_4 solutions including thaumasite in the database (Schmidt et al., 2009)

Researchers, contractors, and owners have been aware of, and monitoring for, external sulfate attack for many years. However, thaumasite sulfate attack is a relatively new discovery in the realm of external sulfate attack and the process of thaumasite formation is not fully understood. Research is ongoing to examine the roles of cement chemistry, exposure temperature, and limestone content on the progression of thaumasite sulfate attack.

2.2.3 Manifestation of Deterioration

The manifestation of external sulfate attack-caused deterioration is a surface attack, causing deterioration of the surface of the element which is in contact with sulfate solution. It is generally accepted that the deterioration front progresses, from the exposure surface into the sample, causing chemical and physical alterations to the hardened cement paste as it progresses. The deterioration front includes a thin (<1.0 mm) damaged region at the surface, underlain by a transition zone (1–3 mm), with an unaltered core region at depth (Schmidt et al., 2009).

Initially, sulfate ions penetrate the concrete's surface and Portlandite is leached from the cement paste. Portlandite ($Ca(OH)_2$) and monosulfate ($C_3A.CaSO_4.12H_2O$) in the hardened cement paste react with external sulfate ions, forming and precipitating gypsum ($CaSO_4.2H_2O$), ettringite ($C_3A.3CaSO_4.32H_2O$) or both. Gypsum and ettringite are expansive compounds; their formation leads to expansion and cracking of the hardened cement paste (Rendell et al., 2002). The deterioration of the paste in this damaged region permits the continued ingress of sulfates, and the progression of the deterioration front, into the concrete.

Underlying this heavily cracked region is a transition zone with fewer and smaller cracks. Initially the overall length change is restrained by the unaltered core; however, as the deterioration front progresses into the sample the core can no longer restrain the expansion and progressive mechanical failure will occur.

Following the reaction of available alumina to form ettringite, thaumasite ($CaSiO_3.CaCO_3.CaSO_4.15H_2O$) formation occurs at the deterioration front. Thaumasite formation occurs due to the interaction between calcium silica hydrates, calcium carbonate, and sulfate ions. The silica ions for thaumasite formation are supplied by the decomposition of calcium silica hydrates. This decomposition causes a loss of integrity of the cement paste and a subsequent loss of material and decrease in mass (Irassar et al., 2005). As the deterioration front progresses into the concrete, causing cracking and chemical alteration of the cement paste, the formation of thaumasite and the loss of integrity follow. Thaumasite is generally only observed at the surface of a sample, where the microstructure has already been seriously damaged by classical (ettringite and gypsum) external sulfate attack-induced cracking. Thaumasite formation is the final stage of external sulfate attack.

2.2.4 Exploring Correlations (Synergies) Between Chemical and Physical Attacks

Damages caused by the crystallization of sodium sulfate salts are often referred to as "physical sulfate attacks" (see previous sections). *Physical sulfate attack* occurs for example in the case of strong and long periods of drying in arid zones such as Western USA, Canada, Australia, Middle East of China, and south of Spain. In this case, significant damage is observed (Zhutovsky & Hooton, 2017). In addition, accelerated and enhanced degradation of the hydration products, as a result of drying and wetting cycles or of cyclic temperature changes, has been reported in the literature (see, e.g., Bassuoni & Nehdi, 2009; Gao et al., 2013), compared to the degradations induced by ESA alone. Wetting and drying cycles may affect concrete structures not only in tidal or splash zones in marine environments, but also in foundations and buried structures, since the level of water in soils, rivers, and sea is likely to vary and consequently the degree of water saturation of concrete exposed to sulfates in these environments will also vary.

In these cases, concrete is exposed to sulfate penetration and to drying (or drying-wetting cycles). Drying of concrete may induce crystallization of salts and possible damage, according to the theory of the crystallization pressure (Scherer, 2004; Flatt, 2002). Concrete is thus submitted to chemical form (induced by ettringite formation) + physical form (induced by salt crystallization) of sulfate attack. In addition, hindered-shrinkage induced cracking in the near-surface zone may enhance the damage.

With regard to the physical attack, wetting and drying cycles or cooling-heating cycles induce repeated dissolution and recrystallization and phase transitions of sulfate salts, which yield volume changes and thus fatigue in specimens/structures. Stresses sufficiently high to overcome the tensile strength of the material may be developed and the salts may exert a sufficient force to bring about disintegration of the material (Bonnell & Nottage, 1939) and to result in progressive spalling/scaling of the concrete near-surface zone (Scherer, 2004; Nehdi et al., 2014; Zhutovsky & Hooton, 2017). Likewise, evaporation was confirmed as a key mechanism for the thaumasite-form of ESA (Mittermayr et al., 2013).

In foundations and structures in tidal zones, both capillary rise and evaporation can occur, which promote supersaturation with respect to mirabilite of the pore solution and high crystallization pressure may occur even in large pores, and crystal efflorescence will occur in this case at the concrete surface. Crystal growth below the material surface (in the near-surface zone pores) or "subflorescences" (see previous sections) may be critical if the crystallization pressure exceeds the tensile strength of the material (Scherer, 2004). The buildup of salt crystallization pressure in the zones of structures which are not submerged may yield poor performance of concrete containing high amounts of GGBS (e.g., 85%) (Gruyaert et al., 2012).

Hence, the most rapid and aggressive form of attack occurs by wick action (rapid ingress of sodium and sulfate ions with the liquid phase inducing crystal growth, followed by crystallization of salts in pores and (hindered) expansion due to phase changes (induced by temperature or RH fluctuations) (Zhutovsky & Hooton, 2017).

2.2.5 *Influence of Sulfate Attack on Corrosion of Rebars*

Concrete reinforcements can corrode if the concrete lowers its alkalinity due to the penetration of carbon dioxide or by the presence of chloride ions. The corrosive effect of the chlorides has been extensively studied; however, other depassivating ions such as sulfates, are much less characterized. Similar to the case of chloride induced corrosion, literature indicates that sulfate ions are able to induce localized or crevice corrosion (Dehwah et al., 2002; Baghabra Al-Amoudi, 1998; Guojian et al., 2016; Shaheen & Pradhan, 2017; Ai-Tayyib et al., 1988; Abd El Haleem et al., 2013; Jarrah et al., 1995; Krishna Vigneshwaran et al., 2016; Argiz et al., 2020). In addition, sulfates have been identified as more aggressive promoters than chlorides of stress corrosion cracking of wires for prestressing (Acha et al., 1990; Krishna Vigneshwaran et al., 2016).

These studies however differ on the value of concentration which could generate such corrosion. One of the reasons of the differences is the several testing conditions. The pH value is seen to influence the threshold, which has been found to vary between 0.00001 and 0.1 M (Guojian et al., 2016; Shaheen & Pradhan, 2017; Abd El Haleem et al. 2013; Ai-Tayyib et al., 1988).

Papers studying the sulfate threshold when present in concrete mix have not been found, except those of (Argiz et al., 2020) that found that an addition of 1% by weight of cement standardized mortar (w/c = 0.5 and cement/sand ratio of 1/3) was enough to develop corrosion as indicated in Fig. 2.16. Related to grouts for post-tensioned structures, Lasa and coworkers in Florida (Krishna Vigneshwaran et al., 2016) have concluded that: "*[SO_4^{2-}/OH^-] = 0.15 may be considered as a conservative limit. This range corresponds to 600 ppm to 3000 ppm of sulfates in pH range 12.6 to 13.3 in pore water solution. (The corresponding samples showing the conservative values had sulfate concentration 0.0007 g/g)*".

The amount of sulfates in the pore solution is relatively high (as much as 0.2–0.3 M) (Page & Vennesland, 1983; Moragues et al. 1987) in the immediate period after water mixing, but the concentration reduces due to the formation of ettringite and monosulfate. The presence of sulfates in the initial hours during setting is usually neglected in the corrosion studies because it is assumed that due to the passivation state of the bars, the sulfate concentration is below the threshold. However, this is doubtful, as it has been detected that corrosion rate values just after mixing that are above the threshold of 0.1 μA/cm^2 in those very early hydration times. This higher corrosion rate can also be because the pH value is not yet alkaline enough or due to the kinetics of formation of the passive layer (Andrade et al., 1986). This can be noticed in the reference specimen of Fig. 2.16. This corrosion rate decreases as the hydration progresses and passivation is achieved but, by coincidence, when the concentration of sulfates in the pore solution is negligible. When the sulfoaluminate phases are formed, sulfate ions are depleted in the pore solution. These sulfates can however be again in ionic state when carbonation displaces the sulfates from the aluminate phases (De Weerdt et al., 2019); however, presence of sulfates is usually neglected when interpreting the corrosion in carbonated solutions (Andrade et al., 1986).

Higher amounts of sulfates can penetrate from outside when the concrete is in contact with contaminated groundwater as well as seawater. In the case of seawater attack the possible effect of the sulfates is shadowed by the higher diffusivity of the chlorides. The studies published on the effect of sulfates alone, either in mortar specimens or in concrete indicate that sulfate ions can corrode both locally and

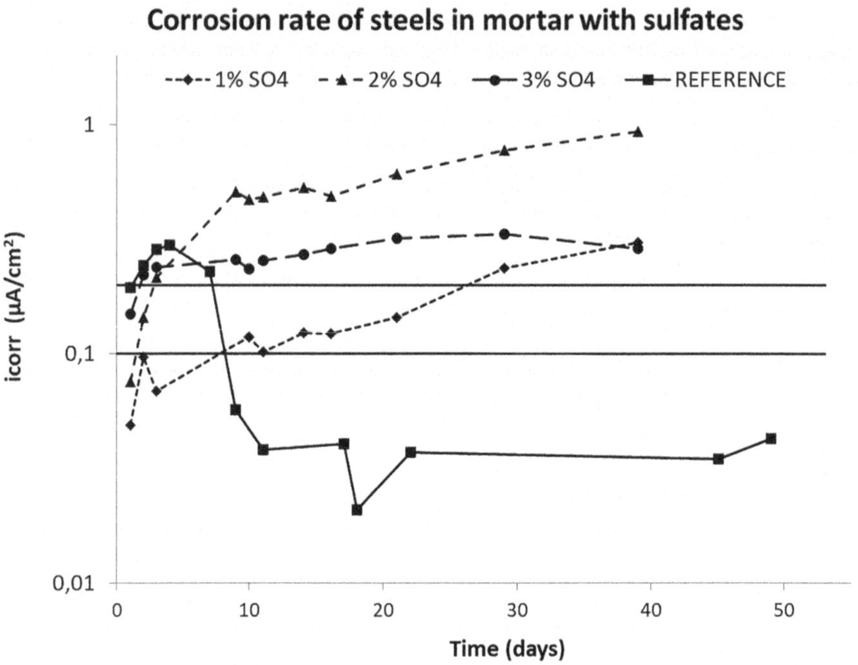

(a)

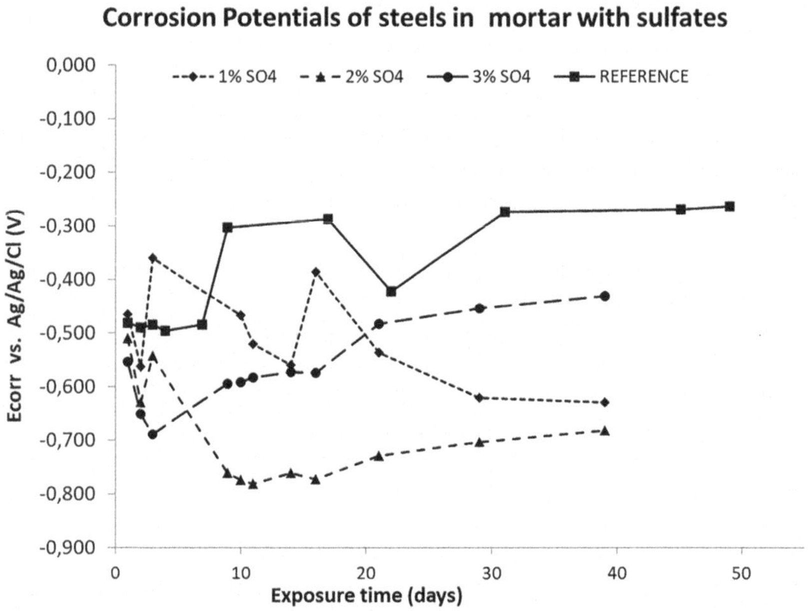

(b)

Fig. 2.16 (**a**) Evolution of corrosion rate in mortar specimens containing sulfates. (**b**) Corrosion potentials. Addition of sulfates from 1% by weight of cement depassivates the steel as indicated by corrosion rates above 0.1–0.2 µA/cm^2

generally with corrosion rates up to similar to those of chloride ions. The amount which depassivates has not been found to be analyzed although, as is the case of chlorides, it likely will be variable depending on the particular concrete and conditions. When the sulfates penetrate with the chlorides, the corrosion rates are reported to be higher than that of the chloride alone (Baghabra Al-Amoudi, 1998; Jarrah et al., 1995).

2.3 Summary

The review of mechanisms of sulfate attack presented in this chapter indicates a complex interplay of physical and chemical processes. While much of the understanding of sulfate deterioration centers around formation of ettringite (and gypsum) and related expansion, the associated changes in the physical states of the materials, as well as the possibility of additional reaction products, alters the scenario significantly. Nevertheless, it is attempted in this compendium to bring about an understanding of the principal mechanisms. In Chap. 4 of this STAR, the focus will be on testing, which will draw upon the concepts discussed in this review.

References

Abd El Haleem, S. M., Abd El Wanees, S. I., & Bahgat, A. (2013). Environmental factors affecting the corrosion behaviour of reinforcing steel. V. Role of chloride and sulphate ions in the corrosion of reinforcing steel in saturated Ca(OH)2 solutions. *Corrosion Science, 75*, 1–15.

Acha, M., Alonso, C., & Andrade, C. (1990). Corrosión bajo tensión de alambres de acero de pretensado en medios alcalinos, conteniendo sulfatos (in Spanish). *Hormigón y Acero, 2*(Trimestre), 89–94.

Adenot, F., & Buil, M. (1992). Modelling of the corrosion of the cement paste by deionized water. *Cement and Concrete Research, 22*(2–3), 489–496.

Ai-Tayyib, A. J., Somuah, S. K., Boah, J. K., Leblanc, P., & Al-Mana, A. I. (1988). Laboratory study on the effect of sulphate ions on rebar corrosion. *Cement and Concrete Research, 18*, 774–782.

Al-Akhras, N. M. (2006). Durability of metakaolin concrete to sulphate attack. *Cement and Concrete Research, 36*(9), 1727–1734.

Al-Amoudi, O. S. B. (2002). Attack on plain and blended cements exposed to aggressive sulphate environments. *Cement and Concrete Composites, 24*(3), 305–316.

Al-Dulaijan, S. U., Maslehuddin, M., Al-Zahrani, M. M., Sharif, A. M., Shameem, M., & Ibrahim, M. (2003). Sulphate resistance of plain and blended cements exposed to varying concentrations of sodium sulphate. *Cement and Concrete Composites, 25*(4–5), 429–437.

Alexander, M. G., Santhanam, M., & Ballim, Y. (2011). Durability design and specification for concrete structures – The way forward. *International Journal of Advanced Engineering, Science and Applied Mathematics, 2*(3), 95–105.

Andrade, C., Alonso, C., Santos, P., & Macías, A. (1986). Corrosion behaviour of steel during accelerated carbonation of solutions which simulate the pore concrete solution. In *8th Congress of Cemnet Chemistry* (Paper 4.3). Rio de Janeiro, Brazil.

Argiz, C., Andrade, C., Moragues, A., Camelli, P., & Bolzoni, F. (2020). Effect of sulphates and sulfides in the corrosion of rebars in mortar. In *Proceedings of Eurocorr 2020*.

Baghabra Al-Amoudi, O. S. (1998). Sulphate attack and reinforcement corrosion in plain and blended cements exposed to sulphate environments. *Building and Environment, 33*(I), 53–61.

Barker, A. P., & Hobbs, D. W. (1999). Performance of Portland limestone cements in mortar prisms immersed in sulfate solutions at 5°C. *Cement and Concrete Composites, 21*, 129–137.

Basheer, L., Kropp, J., & Cleland, D. J. (2001). Assessment of the durability of concrete from its permeation properties: A review. *Construction and Building Materials, 15*, 93–103.

Bassuoni, M. T., & Nehdi, M. L. (2009). Durability of self-consolidating concrete to sulphate attack under combined cyclic environments and flexural loading. *Cement and Concrete Research, 39*(3), 206–226.

Bensted, J. (1999). Thaumasite – Background and nature in deterioration of cements, mortars and concretes. *Cement and Concrete Composites, 21*, 117–121.

Biczok, I. (1967). *Concrete corrosion and concrete protection*. Chemical Publishing, 543p.

Bonen, D. (1992). Composition and appearance of magnesium silicate hydrate and its relation to deterioration of cement based materials. *Journal of the American Ceramic Society, 75*(10), 2904–2906.

Bonen, D., & Cohen, M. D. (1992a). Magnesium sulphate attack on Portland cement paste-I. Microstructural analysis. *Cement and Concrete Research, 22*(1), 169–180.

Bonen, D., & Cohen, M. D. (1992b). Magnesium sulphate attack on Portland cement paste-II. Chemical and mineralogical analyses. *Cement and Concrete Research, 22*(4), 707–718.

Bonnell, D. G. R., & Nottage, M. E. (1939). Studies in porous materials with special reference to building materials: I. The crystallisation of salts in porous materials. *Journal of the Society of Chemical Industry, 58*, 16–21.

Brameshuber, W., Pierkes, R., & Wiens, U. (1999). *Untersuchungen zum Langzeitverhalten von flugaschehaltigen Mörteln Sulfatlösung* (Forschungsbericht Nr. F 729). Institut für Bauforschung.

Brameshuber, W., Pierkes, R., & Wiens, U. (2000). Einfluß der Randbedingungen auf das Prüfverfahren zum Nachweis des hohen Sulfatwiderstandes von Zement-/Flugasche-Gemischen. *Kurzberichte Aus Der Bauforschung, 41*(4), 372–375, Forschungsbericht Nr. F 688. Institut für Bauforschung.

Brown, P. W., & Taylor, H. F. W. (1999). The role of ettringite in external sulfate attack. In J. Skalny & J. Marchand (Eds.), *Material science of concrete – Sulfate attack mechanisms* (pp. 73–98). American Ceramic Society.

Buenfeld, N. R., Shurafa-Daoudi, M.-T., & McLoughlin, I. M. (1995, October). Chloride transport due to wick action in concrete. In *1st RILEM workshop on Chloride penetration into concrete* (pp. 315–324). St. Rémy lès Chevreuse, France.

Burton, M. W., & Howes, D. (1998). *The Sulphate resistance of concretes made with ordinary Portland cement, sulphate resisting cement and ordinary Portland cement and Pozzolan. Report*. United Kingdom Quality Ash Association.

Cao, H. T., Bucea, L., Ray, A., & Yozghatian, S. (1997). The effect of cement composition and pH of environment on sulphate resistance of Portland cements and blended cements. *Cement and Concrete Composites, 19*(2), 161–171.

Chou, I.-M., & Seal, R. R. I. I. (2007). Magnesium and calcium sulfate stabilities and the water budget of Mars. *Journal of Geophysical Research: Planets, 112*, E11004, 10p.

Cohen, M. D. (1983). Theories of expansion in sulfoaluminate – Type expansive cements: Schools of thought. *Cement and Concrete Research, 13*(6), 809–818.

Cohen, M. D., & Bentur, A. (1988). Durability of Portland cement – Silica fume pastes in magnesium sulfate and sodium sulfate solutions. *ACI Materials Journal, 85*, 148–157.

Cohen, M. D., & Mather, B. (1991). Sulfate attack on concrete – Research needs. *ACI Materials Journal, 88*(1), 62–69.

Correns, C. W. (1949). Growth and dissolution of crystals under linear pressure. *Discussions of the Faraday Society, 5*, 267–271.

Cox, W. P., Hornung, E. W., & Giauque, W. F. (1955). The spontaneous transformation from macrocrystalline to microcrystalline phases at low temperatures. The heat capacity of MgSO4·6H2O. *Journal of the American Chemical Society, 77*, 3935–3938.

De Weerdt, K., Plusquellec, G., Belda Revert, A., Geiker, M. R., & Lothenbach, B. (2019). Effect of carbonation on the pore solution of mortar. *Cement and Concrete Research, 118*, 38–56.

Dehwah, H. A. F., Maslehuddin, M., & Austin, S. A. (2002). Long-term effect of sulphate ions and associated cation type on chloride-induced reinforcement corrosion in Portland cement concretes. *Cement & Concrete Composites, 24*, 17–25.

DeKock, C. W. (1986). *Thermodynamic properties of selected metal sulfates and their hydrates* (U.S. Bureau of Mines Information Circular 9081, p. 59p). U.S. Bureau of Mines.

Dhole, R., Thomas, M., Folliard, K., & Drimalas, T. (2011). Sulphate resistance of mortar mixtures of high-calcium fly ashes and other Pozzolans. *ACI Materials Journal, 108*(6), 645–654.

Dhole, R., Thomas, M., Folliard, K., & Drimalas, T. (2013). Characterization of fly ashes for Sulphate resistance. *ACI Materials Journal, 110*(2), 645–654.

Diamond, S., & Lee, R. J. (1999). Microstructural alterations associated with sulfate attack in permeable concretes. In J. Skalny & J. Marchand (Eds.), *Material science of concrete – Sulfate attack mechanisms* (pp. 123–173). American Ceramic Society.

Dikeou, J. T. (1970). *Fly ash increases resistance of concrete to sulphate attack* (Research Report No. 23, A Water Resources Technical Publication, pp. 1–17). Bureau of Reclamation.

Drimalas, T., et al. (2011). *Laboratory and field evaluations of external sulphate attack in concrete*. The National Academies of Sciences, Engineering, and Medicine.

Dunstan, E. R. (1980). A possible method for identifying fly ashes that will improve the sulphate resistance of concretes. *Cement, Concrete, and Aggregates, 2*(1), 20–30.

Dunstan, E. R. (1987). Durability of fly ash concrete. In *Proceedings of the Utilization of Ash Workshop, May 13–15, 1987, III. Fly Ash in Concrete*. University of North Dakota, North Dakota.

El-hachem, R., Roziere, E., Grondin, F., & Loukili, A. (2012). Multi-criteria analysis of the mechanism of degradation of Portland cement based mortars exposed to external sulphate attack. *Cement and Concrete Research, 42*(10), 1327–1335.

Erlin, B., & Stark, D. C. (1966). Identification and occurrence of thaumasite in concrete. *Highway Research Record, 113*, 108–113.

Fernandez-Altable, V. (2009). *Availability of Al2O3 in slag blended cements: Sulphate attack implications*.

Flatt, R. J. (2002). Salt damage in porous materials: How high supersaturations are generated. *Journal of Crystal Growth, 242*(3), 435–454.

Flatt, R. J., & Scherer, G. W. (2008). Thermodynamics of crystallization stresses in DEF. *Cement and Concrete Research, 38*(3), 325–336.

Fraay, A. L. A., Reigersman, A., & De Pee, J. (1987). Sulphate resistance of mortars with pulverized fuel ash. In *Concrete durability – Katharine and Bryant Mather international conference, Atlanta/Georgia, 27 April–1 May 1987* (ACI SP-100, Vol. 2, pp. 2041–2058). American Concrete Institute, Detroit, Michigan.

Gabrisova, A., Havlica, J., & Sahu, S. (1991). Stability of calcium sulphoaluminate hydrates in water solutions with various pH values. *Cement and Concrete Research, 21*(6), 1023–1027.

Gao, J., Yu, Z., Song, L., Wang, T., & Wei, S. (2013). Durability of concrete exposed to sulphate attack under flexural loading and drying-wetting cycles. *Construction and Building Materials, 39*, 33–38.

Glasser, F. P. (1999). Reactions between cement paste components and sulfate ions. In J. Skalny & J. Marchand (Eds.), *Material science of concrete – Sulfate attack mechanisms* (pp. 99–122). American Ceramic Society.

Gollop, R., & Taylor, H. (1995). Microstructural and microanalytical studies of sulphate attack iii. Sulphate-resisting Portland cement: Reactions with sodium and magnesium sulphate solutions. *Cement and Concrete Research, 25*(7), 1581–1590.

Gonzalez, M., & Irassar, E. (1997). Ettringite formation in low C_3A Portland cement exposed to sodium sulphate solution. *Cement and Concrete Research, 27*(7), 1061–1072.

Grevel, K., & Majzlan, J. (2009). Internally consistent thermodynamic data for magnesium sulphate hydrates. *Geochimica et Cosmochimica Acta, 73*(22), 6805–6815.

Gruyaert, E., van den Heede, P., Maes, M., & De Belie, N. (2012). Investigation of the influence of blast-furnace slag on the resistance of concrete against organic acid or sulphate attack by means of accelerated degradation tests. *Cement and Concrete Research, 42*, 173–185.

Guojian, L., Zhang, Y., Ni, Z., & Huang, R. (2016). Corrosion behavior of steel submitted to chloride and sulphate ions in simulated concrete pore solution. *Construction and Building Materials, 115*, 1–5.

Gurevich, V. M., Kuskov, O. L., Gavrichev, K. S., & Tyurin, A. V. (2007). Heat capacity and thermodynamic functions of epsomite MgSO4·7 H2O at 0–303 K. *Geochemistry International, 45*, 206–209.

Hansen, W. C. (1963). Crystal growth as a source of expansion in Portland cement concrete. *Proceedings ASTM, 63*, 932–945.

Hansen, W., & Pressler, E. (1947). Solubility of $Ca(OH)_2$ and $CaSO_4.2H_2O$ in dilute alkali solutions. *Industrial & Engineering Chemistry, 39*(10), 1280–1282.

Hartshorn, S. A., Sharp, J. H., & Swamy, R. N. (1999). Thaumasite formation in Portland-limestone cement pastes. *Cement and Concrete Research, 29*, 1331–1340.

Hime, W. G., & Mather, B. (1999). “Sulphate attack,” or is it? *Cement and Concrete Research, 29*(5), 789–791.

Irassar, E. F. (2009). Sulphate attack on cementitious materials containing limestone filler – A review. *Cement and Concrete Research, 39*(3), 241–254.

Irassar, F., & Batic, O. (1989). Effects of low calcium fly ash on sulfate resistance of OPC cement. *Cement and Concrete Research, 19*(2), 194–202.

Irassar, E.F., V.L. Bonavetti, M.A. Trezza, AND M.A. Gonalez. Thaumasite formation in limestone filler cements exposed to sodium sulphate solution at 20C. Cement and Concrete Composites, 2005, vol. 27, no. 1 pp. 77–84.

Jarrah, N. R., Baghabra Al-Amoudi, O. S., Maslehuddin, M., Ashiru, O. A., & Al-Mana, A. I. (1995). Electrochemical behaviour blended cement concretes in sulphate and/or chloride environments. *Construction and Building Materials, 9*(2), 97–103.

Kamali, S., Gerard, B., & Moranville, M. (2003). Modelling the leaching kinetics of cement-based materials – Influence of materials and environment. *Cement and Concrete Composites, 25*(4), 451–458.

Khatri, R., Sirivivatnanon, V., & Yang, J. (1997). Role of permeability in sulphate attack. *Cement and Concrete Research, 27*(8), 1179–1189.

Ko, H. C., & Daut, G. E. (1979). *Enthalpies of formation of α- and β-magnesium sulfate and magnesium sulfate monohydrate* (U.S. Bureau of Mines Report of Investigations 8409, p. 8p). U.S. Bureau of Mines.

Kohler, S., Heinz, D., & Urbonas, L. (2006). Effect of ettringite on thaumasite formation. *Cement and Concrete Research, 36*(4), 697–706.

Krishna Vigneshwaran, K. K., Permeh, S., Lau, K., & Lasa, I. (2016). *Sulphate limits for corrosion of steel strand in deficient grout.* PTI (Post Tensioning Institute) Convention.

Kunther, W., Lothenbach, B., & Scrivener, K. L. (2013). On the relevance of volume increase for the length changes of mortar bars in sulphate solutions. *Cement and Concrete Research, 46*(4), 23–29.

Kunther, W., Lothenbach, B., & Skibsted, J. (2015). Influence of the Ca/Si ratio of the C–S–H phase on the interaction with sulfate ions and its impact on the ettringite crystallization pressure. *Cement and Concrete Research, 69*, 37–49.

Lang, E. (1993). Sulfatwiderstand von Zement: Einige prüftechnische Aspekte. *Beton-Informationen, 33*(4), 44–47.

Lea, F. M., & Nurse, R. W. (1949). Problems of crystal growth in building materials. *Crystal Growth – Discussions of the Faraday Society, 5*, 345–352.

Lee, B. Y., & Kurtis, K. E. (2017). Effect of pore structure on salt crystallization damage of cement-based materials: Consideration of w/b and nanoparticle use. *Cement and Concrete Research, 98*, 61–70.

Lee, S. T., Moon, H. Y., & Swamy, R. N. (2005). Sulphate attack and role of silica fume in resisting strength loss. *Cement and Concrete Composites, 27*(1), 65–76.

Lipus, K., & Baetzner, S. (2009). Einfluss der Hüttensandchemie auf den Sulfatwiderstand. In *Tagungsbericht, 17. Internationale Baustofftagung* (pp. 2-0311–2-0316). Bauhaus-Universität Weimar.

Lothenbach, B., B. Bary, P. Le Bescop, T. Schmidt, And N. Leterrier. Sulphate ingress in Portland cement. Cement and Concrete Research, 2010, vol. 40. pp. 1211–1225.

Ludwig, H.-M., Muller, M., & Nobst, P. (2012). *Ergänzende Untersuchungen zum DAfStb-Verbundforschungsvorhaben "Vertiefte Untersuchungen zum Sulfatwiderstand von Beton, V 469/2: Abschlussbericht"*. Press: Schriftenreihe des Deutschen Ausschusses für Stahlbeton.

Mangat, P., & Khaitb, J. (1992). Influence of initial curing on sulphate resistance of blended cement concrete. *Cement and Concrete Research, 22*(6), 1089–1100.

Mangat, P., & Khatib, J. (1996). Influence of fly ash, silica fume and slag on sulphate resistance of concrete.

Matthews, J. D. (1991). *The sulphate resistance of PFA concrete* (Note N/91). Building Research Establishment (BRE).

Mcmahan, D. J., Sandberg, P., Folliard, K., & Mehta, P. (1992). *Deterioration mechanisms of sodium sulphate*. Laboratório Nacional de Engenharia Civil.

Mehta, P. K. (1978). Influence of different crystalline forms of C3A on sulfate resistance of Portland cement. *Cement and Concrete Research, 8*, 575–579.

Mehta, P. K. (1983). Mechanism of expansion associated with ettringite formation. *Cement and Concrete Research, 3*(1), 401–406.

Mehta, P. K., Pirtz, D., & Polivka, M. (1979). Properties of alite cements. *Cement and Concrete Research, 9*, 439–450.

Menendez, E., Matschei, T., & Glasser, F. P. (2013). *Sulphate attack in concrete, performance of cement-based materials in aggressive aqueous environments* (pp. 7–74). Springer.

Menendez, E., Garcia–Roves, R., Aldea, B., Ruiz, S., & Baroghel-Bouny, V. (2019). Combination of immersion and semi-immersion tests to evaluate concretes manufactured with sulphate resisting cements. *Journal of Sustainable Cement-Based Material, 8*(6), 337–352.

Messad, S., Carcasses, M., Linger, L., & Boutillon, L. (2010, May 29–June 2). Performance approach using a accelerated test method for external sulphate attack. In *Third International fib Conference, Incorporating PCI Annual Convention Bridge Conference, Washington*. fib-Group, Washington.

Mittermayr, F., Baldermann, A., Kurta, C., Rinder, T., Dietmar Klammer, Leis, A., Tritthart, J., & Martin Dietzel. (2013). Evaporation – A key mechanism for the thaumasite form of sulphate attack. *Cement and Concrete Research, 49*, 55–64.

Monteiro, P. J., & Kurtis, K. E. (2003). Time to failure for concrete exposed to severe sulphate attack. *Cement and Concrete Research, 33*(7), 987–993.

Moragues, A., Macias, A., & Andrade, C. (1987). Equilibrium of the chemical composition of the concrete pore solution. Part 1: Comparative study of synthetic and extracted solutions. *Cement and Concrete Research, 17*, 173–182.

Mulenga, D. M. (2002). *Zum Sulfatangriff auf Beton und Mörtel einschließlich der Thaumasitbildung*. Dissertation, Weimar, Bauhaus-Universität, Fakultät Bauingenieurwesen.

Muller-Rochholz, J. F. W. (1987). *Einfluss von Flugasche, Kalksteinmehl und Kreidemehl auf Festigkeit und Sulfatbeständigkeit von Mörteln. Schlussbericht*. Fachhochschule Münster, Labor für Baustoffe.

Nehdi, M. L., Suleiman, A. R., & Soliman, A. M. (2014). Investigation of concrete exposed to dual sulphate attack. *Cement and Concrete Research, 64*, 42–53.

Nobst, P., & Stark, J. (2003). Investigations on the influence of cement type on thaumasite formation. *Cement and Concrete Composites, 25*(8), 899–906.

Odler, I., & Colan-Subauste, J. (1999). Investigations on cement expansion associated with ettringite formation. *Cement and Concrete Research, 29*(5), 731–735.
Odler, I., & Gasser, M. (1988). Mechanism of sulfate expansion in hydrated Portland cement. *Journal of the American Ceramic Society, 71*(11), 1015–1020.
Pabalan, R. T., & Pitzer, K. S. (1987). Thermodynamics of concentrated electrolyte mixtures and the prediction of mineral solubilities to high temperatures for mixtures in the system Na-K-Mg-Cl-SO4-OH-H2O. *Geochimica et Cosmochimica Acta, 51*(9), 2429–2443.
Page, C. L., & Vennesland, O. (1983). Pore solution composition and chloride binding capacity of silica-fume cement pastes. *Materiaux et Constructions Rilem, 16*(91), 19–25.
Ping, X., & Beaudoin, J. J. (1992). Mechanism of sulphate expansion I. Thermodynamic principle of crystallization pressure. *Cement and Concrete Research, 22*, 631–640.
Planel, D., et al. (2006). Long-term performance of cement paste during combined calcium leaching–sulphate attack: Kinetics and size effect. *Cement and Concrete Research, 36*(1), 137–143.
Ramyar, K., & Inan, G. (2007). Sodium sulphate attack on plain and blended cements. *Building and Environment, 42*(3), 1368–1372.
Rebel, B., Detwiler, R. J., Gebler, S. H., & Hooton, R. D. (2005). The right sulphate test makes a difference. *Concrete International, 27*(02), 49–52.
Rendell, F., Jauberthie, R., & Granthan, M. (2002). *Deteriorated concrete inspection and physicochemical analysis*. Thomas Telford Publishing.
Revertegat, E., Richet, C., & Gegout, P. (1992). Effect of pH on the durability of cement pastes. *Cement and Concrete Research, 22*(2–3), 259–272.
Rodriguez-Navarro, C., Dohne, E., & Sebastian, E. (2000). How does sodium sulphate crystallize? Implications for the decay and testing of building materials. *Cement and Concrete Research, 30*(10), 1527–1534.
Roziere, E., & Loukili, A. (2011). Performance-based assessment of concrete resistance to leaching. *Cement and Concrete Composites, 33*(4), 451–456.
Safiuddin, M., & Hearn, N. (2005). Comparison of ASTM saturation techniques for measuring the permeable porosity of concrete. *Cement and Concrete Research, 35*, 1008–1013.
Sahmaran, M., Kasap, O., Duru, K., & Yaman, I. O. (2007). Effects of mix composition and water – Cement ratio on the sulphate resistance of blended cements. *Cement and Concrete Composites, 29*(3), 159–167.
Santhanam, M. (2001). *Studies on sulfate attack – Mechanisms, test methods and modeling*. Ph.D. thesis, Purdue University.
Santhanam, M., Cohen, M. D., & Olek, J. (2003a). Effects of gypsum formation on the performance of cement mortars during external sulphate attack. *Cement and Concrete Research, 33*(3), 325–332.
Santhanam, M., Cohen, M. D., & Olek, J. (2003b). Mechanism of sulphate attack: A fresh look: Part 2. Proposed mechanisms. *Cement and Concrete Research, 33*(3), 341–346.
Scherer, G. W. (1999). Crystallization in pores. *Cement and Concrete Research, 29*(8), 1347–1358.
Scherer, G. W. (2004). Stress from crystallization of salt. *Cement and Concrete Research, 34*(9), 1613–1624.
Schiebl, P., & Hardtl, R. (1992). *Einfluß von Steinkohlenflugasche (SFA) auf den Sulfatwiderstand von Betonen – Abschlußbericht* (Forschungsbericht Nr. F 262). Institut für Bauforschung.
Schmidt, T., Lothenbach, B., Romer, M., Scrivener, K., Rentsch, D., & Figi, R. (2008). A thermodynamic and experimental study of the conditions of thaumasite formation. *Cement and Concrete Research, 38*(3), 337–349.
Schmidt, T., Lothenbach, B., Romer, M., Neuenschwander, J., & Scrivener, K. (2009). Physical and microstructural aspects of sulphate attack on ordinary and limestone blended Portland cements. *Cement and Concrete Research, 39*, 1111–1121.
Schubert, P. (1974). A contribution on the resistance to sulphate solution of cement mortar prisms containing fly ash. In *Proceedings of the VI. International Congress on the Chemistry of Cement, Sec. III, III 3, 16 Seiten*. Moscow.

Schubert, P., & Lühr, H.-P. (1979). Zum Sulfatwiderstand flugaschehaltiger Mörtel und Betone. *Betonwerk Und Fertigteil-Technik, 45*, 177–182.
Schubert, P., Jaegermann, C., & Glitza, H. (1976). *Kurzzeitwarmbehandlung flugaschehaltiger Betone und Sulfatwiderstand flugsschehaltiger Mörtel (Teile a + b)* (Forschungsbericht Nr. F 72). Institut für Bauforschung.
Shaheen, F., & Pradhan, B. (2017). Influence of sulphate ion and associated cation type on steel reinforcement corrosion in concrete powder aqueous solution in the presence of chloride ions. *Cement and Concrete Research, 91*, 73–86.
Skalny, J., Marchand, J., & Odler, I. (2002). *Sulfate attack on concrete*. Routledge, 232p.
Steiger, M., & Asmussen, S. (2008). Crystallization of sodium sulphate phases in porous materials: the phase diagram Na_2SO_4–H_2O and the generation of stress. *Geochimica et Cosmochimica Acta, 72*(17), 4291–4306.
Sumer, M. (2012). Compressive strength and sulphate resistance properties of concretes containing class F and class C fly ashes. *Construction and Building Materials, 34*(September), 531–536.
Thaulow, N., & Sahu, S. (2004). Mechanism of concrete deterioration due to salt crystallization. *Materials Characterization, 53*(2), 123–127.
Tian, B., & Cohen, M. D. (2000). Expansion of alite paste caused by gypsum formation during sulphate attack. *Journal of Materials in Civil Engineering, 12*(1), 24–25.
Tikalsky, P. J., & Carrasquillo, R. L. (1992). Influence of fly ash on the sulphate resistance of concrete. *ACI Materials Journal, 89*(1), 69–75.
Torill, K., & Kawamura, M. (1994). Effects of fly ash and silica fume on the resistance of mortar to sulfuric acid and sulphate attack. *Cement and Concrete Research, 24*(2), 361–370.
Torill, K., Kawamura, H., Tanaguchi, H., & Hasaba, S. (1987). Influence of fly ash on the resistibility of concrete against sulphate attack. *Transaction Japan Concrete Institute, 9*, 79–86.
Torill, K., Taniguchi, K., & Kawamura, M. (1995). Sulphate resistance of high fly ash content concrete. *Cement and Concrete Research, 25*(4), 759–768.
Tsui, N., Flatt, R. J., & Scherer, G. W. (2003). Crystallization damage by sodium sulphate. *Journal of Cultural Heritage, 4*(2), 109–115.
Verbeck, G. J. (1967). Field and laboratory studies of the sulphate resistance of concrete. *Portland Cement Association, Research Bulletin, 227*, 114–124.
Verbeck, G. J. (1975). Mechanisms of corrosion of steel in concrete. In *Corrosion of metals in concrete* (ACI Publication SP-49, pp. 21–38). ACI Publication.
Wagman, D. D., et al. (1982). NBS tables of chemical thermodynamic data. *Journal of Physical and Chemical Reference Data, 11*, 1–377.
Wang, W., Zeng, D., Zhou, H., Wu, X., & Yin, X. (2015). Solubility isotherms of gypsum, hemihydrate and anhydrite in the ternary systems $CaSO_4{\cdot}MSO_4{\cdot}H_2O$ (M= Mn, Co, Ni, Cu, Zn) at T = 298.1°K to 373.1°K. *Journal of Chemical and Engineering Data, 60*(10), 3024–3032.
Wassing, W. (2003). Relationship between the chemical reactivity of granulated blastfurnace slags and the mortar standard compressive strength of the blastfurnace cements produced from them. *Cement International, 1*(5), 94–109.
Whittaker, M., & Black, L. (2015). Current knowledge of external sulphate attack. *Advances in Cement Research, 27*, 532–545.
Wischers, G., & Sprung, S. (1990). Verbesserung des Sulfatwiderstands von Beton durch Zusatz von Steinkohlenflugasche, Sachstandsbericht Mai 1989. *Beton, 40*(1), 17–21, (2), 62–66.
Yu, C., Sun, W., & Scrivener, K. (2013). Mechanism of expansion of mortars immersed in sodium sulphate solutions. *Cement and Concrete Research, 43*, 105–111.
Zhutovsky, S., & Hooton, R. D. (2017). Accelerated testing of cementitious materials for resistance to physical sulphate attack. *Construction and Building Materials, 145*, 98–106.

Chapter 3
External Sulfate Attacks in the Field

Esperanza Menéndez Méndez, Véronique Baroghel-Bouny, Ricardo García-Roves, Isabel Martins, Florian Mittermayr, Yury Andres Villagrán-Zaccardi, and Claudio Zega

Abstract In this chapter is presented a review of the most significant information about field structures affected by sulfate attacks and concrete test specimens exposed to natural environment. Experience in different countries/continents (North America, Europe, Asia and Africa) is thus reported, pointing out the specificities of these countries with regard to various types of sulfate attacks. For example, cases of thaumasite-type degradation have been documented.

On the basis of this review, an analysis has been carried out, in order to identify the most common situations (according to the environment, the type of structures, etc.).

The influence of various parameters (W/b, type of aggressive agent, pH, sulfate concentration, other aggressive conditions (carbonation, chlorides, etc.), temperature, saturation degree, cement type, aggregates and mechanical loads) on sulfate attacks in concrete structures in service has been studied on the basis of the cases observed. Combined effects with other attacks have been highlighted.

E. Menéndez Méndez
IETcc-CSIC, Madrid, Spain

V. Baroghel-Bouny (✉)
University Gustave Eiffel, Marne-la-Vallée, France
e-mail: veronique.baroghel-bouny@univ-eiffel.fr

R. García-Roves
Universidad Politécnica de Madrid, Madrid, Spain

I. Martins
LNEC, Lisbon, Portugal

F. Mittermayr
University of Innsbruck (former TU Graz), Innsbruck, Austria

Y. A. Villagrán-Zaccardi
VITO NV (Flemish Institute for Technological Research), Mol, Belgium

C. Zega
CONICET LEMIT, La Plata, Argentina

V. Baroghel-Bouny et al. (eds.), *External Sulfate Attacks - Mechanisms, Field Behavior and Lab Test Methods*, RILEM State-of-the-Art Reports 48,
https://doi.org/10.1007/978-3-032-19402-2_3

The specific case of soils as an external source of sulfates has been detailed. The exposure classes, with respect to sulfate concentration in soils and groundwater, according to various international standards, are summarized. The methods of mitigation of the effects of ESA and acid attacks are also reported.

Keywords Damage · ESA · Ettringite · Field conditions · Thaumasite · Structures · Soils

3.1 Introduction

The deterioration of concrete by external sulfate attack in field structures has been investigated by some researchers in recent decades, using techniques of visual evaluation, microscopy, mass loss, and analysis of physical and chemical properties of concrete in general. It is not very common to find literature worldwide regarding attack by sulfates in concretes of field structures. Naturally occurring sulfates of sodium, potassium, calcium, or magnesium, that can attack hardened concrete, are sometimes found in soil or dissolved in groundwater adjacent to concrete structures. Sulfate salts in solution enter the concrete and attack the cementing materials. If evaporation takes place from a surface exposed to air, the sulfate ions can concentrate near that surface and increase the potential for causing deterioration.

Sulfate attack of cement-based construction materials has occurred at various locations throughout the world, being a particular problem in arid areas. The two best recognized chemical consequences of sulfate attack on concrete constituents are the formation of ettringite (calcium aluminate trisulfate 32-hydrate, $CaO \times Al_2O_3 \times 3CaSO_4 \times 32H_2O$) and gypsum (calcium sulfate dihydrate, $CaSO_4 \times 2H_2O$). Moreover, it is worth mentioning the formation of thaumasite ($Ca SiO_3 \times CaCO_3 \times CaSO_4 \times 15H_2O$) in low temperature exposure conditions, several times associated with ettringite.

The formation of ettringite can result in an increase in solid volume, leading to expansion and cracking while the formation of gypsum can lead to softening and loss of concrete strength. The presence of ettringite or gypsum in concrete, however, is not in itself an adequate indication of sulfate attack; evidence of sulfate attack should be verified by petrographic and chemical analyses. When the aggressive sulfate solution contains magnesium sulfate, brucite ($Mg(OH)_2$, magnesium hydroxide) is produced in addition to ettringite and gypsum. It should also be pointed out that some of the sulfate related degradation processes can damage concrete without expansion. For example, concrete subjected to soluble sulfates can suffer softening of the paste matrix or an increase in the overall porosity, either of which diminish durability (ACI, 2001).

Throughout the service life of concrete structures, alterations can be produced as a result of environmental factors which could affect their functionality, compromising the safety of the structures or the structural elements, according to the degree of deterioration (Menéndez, 2010; BCA, 1988; Hobbs, 1988; West, 1996; Geoffrey &

Alexander, 2011; Skalny et al., 2002; Scrivener & Skalny, 2002; Thomas & Ramlocham, 2002; Menéndez, 2002; Stutzman, 1999; Jacobsen et al., 2008; Tepponen & Eriksson, 1987; Shayan & Quick, 1992; Oberholster et al., 1992; Menéndez et al., 2011).

This deterioration might be associated with different types of factors: chemical, physical, mechanical, biological, etc. In some cases the origin of the damage is directly related to the characteristics of the concrete constituents (aggregates, cement, water, additives, etc.) and the interactions between them or with the environment.

In thaumasite sulfate attack to concrete structures, the favourable conditions which encourage the formation of thaumasite within cement pastes are very damp, cold environments, available sulfate and carbonate ions and some reactive form of alumina (Crammond, 1985). The following combination of conditions are conducive to thaumasite attack: a low temperature, wet conditions, exposure to sulfates, particularly those associated with the magnesium cation and/or exposure to sulfuric acid, the prior formation of ettringite, the depletion of calcium hydroxide and the inclusion within the concrete of at least 10% to 25% calcium carbonate by mass of the cement (Hobbs, 2003).

The mechanism of degradation of concrete exposed to external sulfate attack depends on exposure conditions, such as temperature (Crammond, 2002a, b; Lipus & Punkte, 2003), associated cation (Duval & Hornain, 1992; Neville, 2004) and sulfate concentration (Santhanam et al., 2001; Higgins & Crammond, 2002; Dehwah, 2007).

In all case of alterations arising from chemical reactions originating internally, water acts as a vehicle for the reaction and as a basic part of the expansive products formed.

Within the expansive processes of an internal origin it is important to highlight, given its impact and the effects produced on the concrete, the alkali-silica reaction and the sulfate attack, although there are other phenomena of a physical origin that have an impact on the freezing and thawing cycles. Naturally both chemical and physical attack can occur in one single specimen (Nehdi et al., 2014).

The alkali-silica reaction and the sulfate attack are phenomena brought about by the chemical reaction of different constituents of the concrete which give rise to the neo-formation of products whose main characteristic is that they are expansive.

The formation of these expansive products in the interior of the hardened concrete makes it overcome the tensile strength of the material and gives rise to the formation of cracks.

Another significant characteristic is that its occurrence is affected by multiple factors inherent to the material itself and its constituents as well as by environmental conditions. The coincidence of influential factors and the manifestation of the damage, with similar characteristics, makes it essential to carry out a suitable diagnosis in order to program actions on the structure and carry out repair work where deemed necessary.

In the event of chemical sulfate attack, macroscopic investigations are marked by expansion, ultimately cracking and/or spalling. Expansion is perhaps the most common form of assessing sulfate attack in the laboratory, as decreed by the standard ASTM C1012-13M (ASTM, 2018). Loss of strength, dynamic modulus and a change in mass is also indicative of damage. Softening and decohesion may also occur. All of these observations are the result of internal changes of the microstructure. Sulfates react with aluminate hydrates to produce ettringite, precipitating from a supersaturated solution; its growth exerts pressure in small pores, causing damage. The role of gypsum is still debated, with it being stabilised when highly concentrated solutions are used. Some studies, however, have noticed the presence of gypsum when lower concentrated solutions are used.

Factors that accelerate or aggravate sulfate attack are as follows (ACI, 2001):

- High porosity due to:
  - High water absorption
  - Permeability
  - Voids
- Cracks and discontinuities due to:
  - Stress concentrations
  - Thermal shock
- Leaching and liquid penetration due to:
  - Flowing liquid (movement of water-carrying deleterious substances increases reactions that depend on both the quantity and velocity of flow)
  - Ponding
  - Hydraulic pressure

And the factors that mitigate or delay attack:

- Dense concrete achieved by:
  - Proper mixture proportioning (the mixture proportions and the initial mixing and processing of fresh concrete determine its homogeneity and density)
  - Reduced unit water content
  - Increased cementitious material content
  - Air entrainment
  - Adequate consolidation
  - Effective curing (poor curing procedures result in flaws and cracks)
- Reduced tensile stress in concrete by (resistance to cracking depends on strength and strain capacity):
  - Using tensile reinforcement of adequate size, correctly located
  - Inclusion of pozzolan (to reduce temperature rise)
  - Provision of adequate number of contraction joints
- Structural design:
  - To minimize areas of contact and turbulence
  - Provision of membranes and proactive-barrier systems to reduce penetration (Concrete that will be frequently exposed to chemicals which are known to

produce rapid deterioration should be protected with a chemically resistant protective-barrier system)

The UK Highways Agency Expert Group had identified primary and secondary factors that must coexist or would influence the occurrence or severity of thaumasite sulfate attack (Loudon, 2003).

- Primary factors:
  - presence of sulfates and/or sulfides in the ground;
  - presence of mobile groundwater;
  - presence of carbonate generally in coarse and/or fine concrete aggregates;
  - low temperatures.
- Secondary factors:
  - type and quantity of cement used;
  - quality of concrete mix and compaction;
  - changes to ground chemistry and water regime resulting from construction;
  - type, depth and geometry of buried concrete.

Typical investigations rely on laboratory studies to assess sulpate resistance. Usually these researches are accelerated, by using strongly concentrated sulfate solutions. This approach, however, often does not necessarily reflect field conditions. This is recognised, and more recent studies change exposure conditions (pH, renewal, concentration) so as to better mimic field conditions.

Recently, attention has been given to physical sulfate attack. Physical sulfate attack does not entails any reaction with the hydrates. It is simply the recrystallisation of sulfate salts in pores as the solution is allowed to dry. Drying can happen if exposure is cyclic or when specimens are semi-immersed. In the latter case, dissolved salts rise through the sample by capillary action, before recrystallising where the humidity drops. There is still, however, no standardised test for this sort of exposure.

The extent of damage is dependent on the nature of the mix. In the case of chemical sulfate attack, the use of supplementary cementitious materials (SCMs), combined with low w/c, will result in improved performance. In the event of physical sulfate attack, the use of SMCs, however, can worsen damage as greater capillary action can arise from a refined pore structure.

In external sulfate attack to structures, sulfate ions can penetrate through the cement matrix, leading to damage. In buried structures, the entering sulfates react with aluminate hydrates to produce ettringite and gypsum. Damage can also be caused by the precipitation of sulfate salts in porous structures such as concrete. This physical form of damage, however, requires some form of drying of the solution allowing sulfates to crystallise in pores. Such a mechanism can be encountered for example in tidal zones, where a cyclic wetting and drying phenomenon can occur, and in half buried structures (Whittaker & Black, 2015).

Concrete can be vulnerable to damage when salt crystals grow from a supersaturated solution in its pores (Scherer, 2004). This process was described as a physical attack on concrete since, unlike the chemical sulfate attack; it does not involve chemical interactions between sulfate ions and the concrete hydration products (Haynes et al., 1996).

The nature of the manifestation of sulfate degradation depends on several factors including the exposure conditions that can range from continuous immersion to cycles of wetting and drying.

If the concrete is essentially continuously immersed in a sulfate solution and sulfate attack occurs, either cracking or delamination of concrete surfaces or their softening, or all three phenomena simultaneously, are usually observed. The affected area or reaction zone follows the advancing sulfate penetration front. These processes result in a reduction of the concrete's mechanical properties.

Wetting and drying cycles with sulfate-containing water is another form of exposure which can be encountered when water levels rise and drop, e.g., by flooding and runoff of precipitation, and ocean tides. Also, cyclic migration of water into concrete can be induced by capillary flow and cyclic variation in atmospheric relative humidity. The result of such processes is the increase of sulfates concentration near the free concrete surfaces. When soluble sulfate salts, as well as other soluble salts, are concentrated in sufficient amounts, disintegrations of the surface layers can occur.

In a restricted sense, the degradation of concrete by concentrated sulfate solutions in the outer concrete surfaces through wetting and drying cycling is a physical process and other soluble salts may also induce the same type of damage. However, if the wetting and drying processes are gradual then it is likely that both chemical and physical processes are involved in the degradation.

Durability of concrete exposed to sulfates has primarily been studied on specimens fully-submerged in sulfate solutions. However, field experience with concrete exposed to sulfates has often shown that concrete mainly suffers from surface scaling caused by physical sulfate attack. Such damage is generally limited to the above-ground portion of concrete, while the portion embedded in the sulfate rich soil, which is directly exposed to chemical sulfate attack, is generally found in better condition. The damage process involves capillary rise and evaporation of the groundwater containing sulfates at the above ground concrete surface, resulting in crystal growth in concrete pores and subsequent damage (Haynes et al., 1996; Irassar et al., 1995; Stark, 1989; Yoshida et al., 2010). Such problems have been observed in basements, foundations, and footings of residential dwellings, and service tunnels and dams (Clifton, 1999). This damage has often been ignored and even misinterpreted as chemical sulfate attack. In laboratory, test specimens partially-immersed in sulfate solutions and exposed to temperature and relative humidity cycles show that concrete can experience dual sulfate attack. Damage occurs in areas exposed to the dry atmosphere in the form of spalling and scalling from both chemical and physical processes. The lower immersed portion can suffer from chemical sulfate attack, while the upper portion can be vulnerable to physical attack (Nehdi et al., 2014).

Moreover, the consequences of physical sulfate attack on concrete are different from that of chemical sulfate attack since it leads to surface degradation similar to that caused by cycles of freezing and thawing. Conversely, chemical sulfate attack usually results in expansion and cracking due to the formation of ettringite and gypsum (Menéndez et al., 2011).

The vulnerability of concretes to damage by physical salt weathering depends on their pore structure. Concretes having higher volume of micro-pores connected with larger pores are the most vulnerable to damage by salt crystallization. The presence of micro-pores increases the capillary rise and the surface area of evaporation, leading to higher supersaturation of the pore solution and subsequently more damage occurs (Nehdi et al., 2014).

Cases of sulfate attack in field structures are fairly uncommon, mainly because most agencies have already adopted specifications and guidelines aimed at preventing deterioration. Even when sulfate attack does occur, it is not clear how damaging the distress can be in the long term but Metha has suggested that the "threat" of structural failure due to sulfate attack is less than that for other distress mechanisms, such as corrosion or alkali-aggregate reaction (Mehta, 2000). One major issue is that there is a general lack of correlation between accelerated laboratory test methods and actual field performance, which is discussed next (Drimalas et al., 2011).

3.2 Field Structures Affected by Sulfate Attack

In order to perform an analysis of structures attacked by sulfates according to different parameters that favour the damage, a literature search of structures damaged by sulfate attack had been carried out and is presented in this section. Here, details of structures with damage caused by sulfate attack worldwide are discussed.

3.2.1 North America

In Canada, one case includes the study of buried concrete and slab in contact with soils. Sections of a concrete aqueduct, built between 1915 and 1919 to supply water to the City of Winnipeg, Manitoba, which were suspected of suffering from classical sulfate attack in areas where the 156-km long structure passes through subsoils with a high sulfate content (Park, 1996). Signs of the thaumasite form of sulfate attack were found when concrete was examined by optical and electron microscopy (Rogers et al., 1997). Petrographic examination revealed abundant deposits of thaumasite, filling continuous sinuous cracks parallel to the outside face of the concrete; these cracks varied from 20 to 120 μm in width and occurred at the edges of coarse aggregate particles and within the matrix (Thomas et al., 2003).

Another investigation in Canada analyses a 33 year-old concrete pavement near Chatham, Ontario, which was suffering from D-line cracking, revealing the presence of thaumasite in concrete cores removed from the westbound lanes (Rogers et al., 1997). The deterioration was attributed to the frost susceptibility of the coarse aggregate, a crushed dolomite. Petrographic examination of the concrete revealed extensive cracking principally in the form of a series of horizontal fractures between 0.1 and 1.0 mm wide and generally consistent with freeze–thaw deterioration (despite the presence of an apparently adequate air void system). Many of the coarse aggregate particles had undergone a de-dolomitization process resulting in carbonation of the surrounding cement paste. Thaumasite was found filling air voids in discrete areas of the concrete and the cement paste in these areas was characterized by a light-coloured mottled appearance. Typically these areas of mottling were about 20–50 mm in size and occurred irregularly throughout the concrete (Thomas et al., 2003).

In Artic, (Bickley et al., 1993a) reported on thaumasite sulfate attack in 4-year old concrete in the Polar Continental Shelf Project facilities at Resolute Bay in the Canadian Artic. The gravel used contained dolomite, calcite and quartz. Columns at the Arctic site were severely distressed at ground level and the authors considered that it was likely that the initially placed concrete was not able to cure sufficiently before being subjected to a combination of freezing temperatures at night and aggressive sulfate/carbonate rich water from the partially thawed surrounding terrain. In a subsequent paper (Bickley et al., 1993b) authors concluded that the existing moisture in the soil samples could present the most severe sulfate exposure conditions for concrete. Paste concentrates prepared from severely distressed parts of the columns showed extensive amounts of gypsum, thaumasite, calcite and quartz and an absence of calcium-silicate-hydrate, portlandite and ettringite. The sulfate contents of the paste concentrates were in the range 22.0–26.4% expressed as SO_4 by mass (Hobbs, 2003).

In 1965 in USA, it was observed thaumasite attack in sanitary sewer pipes and at the base of a core taken from an 11-year old pavement. One pipe was severely deteriorated to a depth of about 12 mm beneath which there were deposits of thaumasite, ettringite, gypsum and traces of calcite (Erlin & Stark, 1965; Stark, 2003). This was the first detected case of concrete deterioration by thaumasite sulfate attack.

Texas transportation structures exposed to gypsiferous soils or gypsum-bearing groundwater were investigated in Drimalas et al. (2011). To pinpoint areas of possible sulfate attack, agricultural maps identifying sulfate-bearing soils, were utilized. The El Paso, Paris, and Childress Districts showed the most promising areas containing sulfate-bearing soils in the state of Texas. Over 200 concrete bridges, culverts, columns, and other transportation structures were visually inspected, and a subset of these were selected for further, detailed investigations. In Childress District, Site 5 consisted of damage to a concrete wall below a bridge on US 83. Site 9 had riprap damage below a bridge. Riprap is a sloped wall of concrete used to hold back earth/soil to prevent erosion. Site 10 was a bridge on FM 2278 that had deteriorating columns. In Paris District, Site 22 was a culvert on Highway 19, north of Sulphur Springs and Site 57 was a bridge showing signs of distress in the columns that were located on State Highway 11 northwest of Sulphur Springs. In El Paso District, three of the locations included riprap damage while the fourth case involves concrete right of way (ROW) markers in the area. Sites 3, 31 and 32 all had riprap damage and Site 27 was the location of the deteriorated ROW markers. In damaged structures deleterious reaction products as ettringite, gypsum, portlandite, and monosulfate were found (Drimalas et al., 2011).

Problems with the deterioration of concrete pavement were noted by the Testing Division of the Mississippi State Highway Department as early as the 1950s. In the 1960s, removal and examination of damaged concrete from pavement that was installed over soils known to contain high levels of sulfate showed mineralogical evidence that calcium alumino-sulfate minerals such as ettringite were present in significant quantities (Lossing, 1966; Weiss, 2010). The areas in Mississippi where concrete deterioration associated with sulfate were observed, and the areas where sulfates have been reported to be abundant were Paviment in Northeastern MS, Tennessee-Tombigbee Waterway; South-southwestern MS-eastern LA, Chickasaw Co., Itawamba Co., Lee Co., Lauderdale Co., Issaquena Co. in Mississippi (Lossing, 1966; Ammons et al., 1991; Aslan & Autin, 1996; Jones, 2000; Weiss, 2010).

Unlike the western states, which are drier, Mississippi does not have evaporative buildup of sulfates in the soil. The sulfate in Mississippi comes from the oxidation of sulfide minerals that are present in the newer peat-rich soils and in the older superficial geologic deposits. The recent deltaic deposits are known to contain finely divided, amorphous iron sulfide from decomposing plant materials. The older geologic sulfides are well-crystalline minerals related to pyrites and marcasites (FeS_2) that formed during the Tertiary marine sediments in the Mississippi Embayment. In some areas of Mississippi, the determination of the amount of sulfate in the soil or groundwater may not be as important in estimating the possible extent of sulfate damage as knowing the amount of potential sulfate production represented by the unoxidized iron sulfides in the soil. Disturbing the soil by excavating to install concrete footings and lay down concrete pavement may be the most important factor in exposing the buried sulfides to air, so they can be oxidized to form acidic sulfate compounds that can aggressively attack the later in-place concrete (Weiss, 2010).

In Southern California (La Palma, Lakewood, and Cypress, CA), concrete foundations of homes were investigated. The concretes examined were obtained from homes in a development in Southern California built on a hillside facing the Pacific Ocean. Prior to construction, the standard practice of cutting and filling to produce the grading required for roads, driveways, footings, and garage slabs was carried out. The homes themselves were built into the hillsides and their foundations are comprised of retaining walls, foundation piers, and stem walls. Sodium, magnesium, sulfate, chloride and carbonate have participated in the formation of a variety of phases in concretes subjected to the action of groundwater. In particular, magnesium silicate, brucite, Friedel's salt, thaumasite and likely some species of sodium carbonate had formed. The concrete specimens studied were obtained from cores which have not been in direct contact with the soil. Thus, the phases described formed as a result of the transport of sodium, magnesium, sulfate, chloride and carbonate to significant distances through the pore structures of these concretes. These concretes were cast with a water-to-cement ratio in the range of 0.65–0.70. The elevated water-to-cement ratio used permitted the ingress of aggressive species and efflorescence was routinely observed on the evaporative surfaces of concrete members (Brown & Doerr, 2000).

Also, microscopy examination of concrete, obtained from various structural locations of residential homes of Southern California, revealed the presence of thaumasite in association with ettringite and/or gypsum. Thaumasite was identified only in few cores obtained from garage slabs. This concrete did not contain ground carbonaceous (limestone, dolomite) material as filler or as aggregates. Severe carbonation was observed at the bottom as well as at the top of the concrete. The depth of carbonation was measured to a 20–30 mm thickness. Though carbonation was observed at the top and bottom of the concrete, thaumasite deposits were observed only in the bottom portion of the cores. The investigation concluded that the potential of thaumasite formation was high when there was adequate carbonates available in addition to the sulfates present in the groundwater. Thaumasite formation was another indicator of concrete distress. The formation of thaumasite consumed the C–S–H binder. The general notion that thaumasite formation from sulfate attack occurs only in cold environments is not true, although thaumasite may form in higher amounts in lower temperatures (Sahu et al., 2002).

Also (Sahu et al., 2003) investigated cores from a garage floor of a residential home. The combined attack of sulfates and carbonates from the groundwater altered the original cement paste in addition to the precipitation of sulfate minerals. The concrete examined did not contain any carbonate bearing aggregates or fillers. The required carbonate and sulfate ions for thaumasite formation were derived from the groundwater. Thaumasite deposits were observed in air voids alone or in association with gypsum and/or ettringite. SEM–EDS analysis showed deposits of ettringite, thaumasite and intermediate phases within the same air voids indicating that ettringite is first formed followed by thaumasite with a series of solid solutions. In this reaction process the pH of the local environment and the balance between sulfate, silicate and carbonate ions dictated the type of mineral assemblages.

Also in Southern California, in cores taken from residential homes from Southern California, thaumasite deposits were present in association with gypsum and ettringite.

In southern U.S.A, premature deterioration of the bridge columns in the form of cracking was observed shortly before the concrete reached 10 years of age. Delayed ettringite formation was confirmed as the sole cause of distress in one cast-in-place concrete column and as a contributory cause in two other columns. The action of damaging delayed ettringite formation was readily established by the presence of numerous aggregate particles completely surrounded by ettringite-filled gaps (Thomas et al., 2008).

3.2.2 Europe

Several cases of thaumasite formation within brickwork constructions had been found in various parts of mainland Britain. Brickwork constructions with Portland cement mortar in South Wales were investigated. The first building investigated was a large building, square, flat roofed construction situated on an industrial site. It comprised a 200 mm concrete inner leaf separated from 105 mm keyed OPC/Fletton brickwork by an expanded polyurethane insulation layer, 50 mm thick in a 75 mm cavity. The whole structure was rendered with a 10 mm thick 1:4 cement:sand (by weight) first coat and a 0.1 mm thick 1:2 second layer. The second coat was made much stronger as it was used as a bedding layer to accommodate large area, glazed ceramic tiles. This building was particularly interesting as fine cracking was initially seen along the horizontal and vertical joints of the external tile layer within a short time period of about 3 months after the date of completion. A year followed which included a very cold winter and the cracking became much more severe with loss of adhesion between render' and brickwork. Cracking within the first 3 months was caused by drying shrinkage of the cement-rich render coats. Once cracks appeared in the render, water can seep through to the interface with the Drickwork and into the brickwork beyond. Because of the impervious render and tile covering, this water become trapped and eventually saturates the brickwork thus generating ideal conditions for both sulfate and frost attack. This two latter phenomena had produced the second phase of cracks which runned in horizontal and vertical lines parallel to the underlying mortar joints. In one particularly damaged wall severe bowing of the render away from the brickwork had resulted in spalling of a small area of render leaving the brickwork exposed (Crammond, 1985).

The second building investigated was an 8 year old detached, executive-style house which was of Fletton brickwork construction, laid and rendered using an OPC mortar. The two render coats were thinner than recommended, the first being 2–5 mm thick and the second between 4 and 10 mm. In exposed elevations, the typical horizontal cracking associated with sulfate attacked rendered brickwork was evident and had firstly been noticed 4 years before the samples were taken. Some of this damage was also probably caused by frost action. In particularly susceptible areas, for example the chimney stacks, the render had actually fallen off to reveal a layer of white powdery material up to 5 mm deep coating the entire brickwork surface. The white material has since been identified as a thaumasite-rich mortar and presumably originated from the action of sulfate ions on the back of the render. As in the case of the first site, the whole process of sulfate attack was initiated through drying shrinkage of the render coats. Again a stronger second render coat, approximately 1:2 cement:sand (by weight) on top of a weaker 1:3 first coat had caused the brickwork/render interface and the Fletton brickwork itself to become saturated with rain water. First of all the render was removed to expose the brickwork underneath and, although the white material was still evident along the interface, this time it had the consistency of a wet, gritty mush. It was interesting to find that although the back of the render had deteriorated to a mush, the weaker brickwork mortar was still sound has shown that it contained some ettringite and thaumasite; enough in fact to generate the horizontal cracks mentioned earlier. The occurrence of both ettringite and thaumaslte together was responsible for expansion of rendered brickwork but it was thaumasite alone which causes severe softening of the cement matrix (Crammond, 1985).

In Abbeymead, Gloucester, United Kingdom, a 9 year old concrete foundation to a detached garage on sloping ground, the pyritic Lower Lias clay in contact with the foundation was disturbed as a consequence of wetting and drying and the presence nearby of a 3.5 m high hawthorn hedge and a 10 m high ash tree (National House Building Council, private communication). The depth to which thaumasite sulfate attack occurred in concrete in contact with disturbed Lower Lias clay, reported by BRE, was up to 50 mm at 9 years (<25 mm of sound concrete) PC strip foundation have a w/c 0.55 (Crammond et al., 1999; Hobbs, 2003).

Also in United Kingdom, concrete piles were damaged by sulfate attack. The possible sources of degradation were the oolitic limestone aggregate and sulfate bearing groundwater. The type of degradation was profusion of thaumasite in 300 mm (Crammond & Halliwell, 1995; Rahman & Bassuoni, 2014).

In 1996, thaumasite sulfate attack in *in-situ* concrete foundation piles below a 4-year-old United Kingdom housing estate was studied. This was the first recorded case where deterioration due to sulfate attack had occurred in concrete foundations specifically designed to provide good sulfate resistance with sulfate-resisting Portland cements. The cement content was found to be in excess of 400 kg/m^3 and the w/c ratio was in the range 0.3–0.5. The piles had been exposed to the Lower Lias Clay and the groundwater flowing on the site was Class 3 (suitable for most uses including, individual water supply, irrigation, agriculture and general industrial and commercial use). Excessive amounts of this water had percolated through the tops of the piles, which had transformed into a white, pulpy mass (BRE, 1996; Crammond, 2002a, b).

Thaumasite was found in the foundations to M5 motorway bridges in the west of England (TEG, 1999; Sibbick & Crammond, 1999). Thaumasite sulfate attack was first observed in early 1998 on a Highways Agency structure in the substructure of the Tredington-Ashchurch Bridge, an overbridge carrying a local road over the M5 Motorway situated between Junctions 9 and 10 in Gloucestershire. At about the same time similar, but less severe defects, were also found in Grove Lane Bridge foundations, also on the M5, but further south, between Junctions 12 and 13. Tredington-Ashchurch is a four span overbridge with a reinforced concrete deck. Supporting columns are sited in the central reserve and verges, and extend through placed fill for about 5 m to reinforced concrete spread footings. The bridge was built about 30 years ago, to the applicable contemporary Ministry of Transport standards and specifications. Regular structural inspections and testing of the concrete elements above ground had not indicated any observable signs of distress to the bridge in the intervening period. Planned strengthening work to the columns of the bridge necessitated excavation of some of the backfill surrounding the concrete supports, and in the course of this work site staff observed some unusual deterioration in the exposed concrete. This indicated that some of the concrete surfaces had turned to a mushy consistency and of warty appearance with evidence of expansion of the residual material. This was unexpected to say the least, and potentially extremely serious. Diagnosis was first made by Maintenance Agents, Gloucestershire County Council/Halcrow and later confirmed by the Building Research Establishment (BRE) as thaumasite sulfate attack, after initial concrete samples had been analysed. A number of critical factors in combination had occurred, which were significant in terms of the deterioration that occurred, and particularly in relation to the severity of the thaumasite sulfate attack: use of limestone aggregates in the concrete; use of sulfate and sulfide bearing materials for backfilling around the foundations and buried columns (the soil excavated to construct the foundations contained iron pyrites (sulfides) which started to oxidise after exposure to the atmosphere, and added to the reservoir of available sulfates); groundwater movements; physical disposition of the structure (deep foundations and slender concrete elements); construction regime, particularly the excavation of a large hole through undisturbed ground which was subsequently backfilled with the excavated materials and created a sump around the buried columns; leaking motorway drainage; relatively cold conditions (Loudon, 2003).

As a result, the United Kingdom Government convened the Thaumasite Expert Group (TEG), who were commissioned to prepare a Report 'The thaumasite form of sulfate attack: Risks, diagnosis, remedial works and guidance on new construction' (Thaumasite Expert Group, 1999; Crammond, 2002a, b).

About 50 new cases of thaumasite sulfate attack had been identified in cementitious components of buildings and structures in the United Kingdom since the publication of the TEG Report and 30 since the publication of the first review in 2000. The majority of these were in the foundations of bridges founded on sulfate-bearing Lower Lias Clay, found as a result of Highways Agency investigations of highway sub-structures in Gloucestershire, Somerset and Wiltshire. The remainder had occurred in a wide range of structures, buildings and environmental conditions (Clark & BRE, 2002). Some cases studies comprise: One case in the Rugby area of oversite concrete above hard-core comprising brick-rubble and plaster demolition debris; In Gloucestershire part of a trench fill foundation to a domestic garage founded on Lower Lias Clay; In County Durham highway bridge sub-structure on the A1 (M) of 32 years; In Gloucestershire/Avon 17 more cases in highway bridge sub-structures founded on the Lower Lias Clay (Clark & BRE, 2000a, b).

Between the new cases of concrete structures with damage by sulfate attack after the review in 2000, two were selected. The first case occurred in buried structural concrete of A66(M) Area 18 (northeaster England), where the possible sources of the sulfate were made ground comprising dark brown clay with occasional fragments of red shale (burnt colliery spoil), and *in-situ* clay containing fragments of coal. Both materials had a significant oxidasable sulfides content with a maximum of 1.61 g/l SO_4 indicating pyrite to be present. Needles of thaumasite within voids, were found to a depth of 3.5 mm.

The second case concerns to two piers constructed with different concretes of A1 (M), also in Area 18. The east pier was manufactured with a Portland cement content of 390 kg/m^3 and a water/cement ratio of 0.56. The west pier was manufactured with a Portland cement content of 330,380 kg/m^3 and a water/cement ratio of 0.58. The source of the sulfates was possibly ground containing occasional to abundant fragments of shale (unburnt colliery spoil). It contained a significant amount of oxidisable sulfide 1.7 g/l SO_4 indicating pyrite to be present. The east pier had a maximum depth of thaumasite sulfate attack, TSA, of 6 mm with no surface softening. The aggregate was limestone coarse and siliceous fine aggregate. The west pier had thaumasite sulfate attack penetration to a depth of 22 mm including a 3 mm softened zone. This was surprising, since the concrete contained aggregate with siliceous gravel coarse and fine fractions.

Other cases were investigated as cases of thaumasite sulfate attack in Highways Agency Area 2 (M5 and M4 in Gloucestershire, Somerset and Wiltshire), in Highways Agency Area 18 (County Durham), cases in sulfate-bearing brickwork, cases in internal rendered walls contaminated with gypsum, a case in harbour steps exposed to seawater, several cases of floor slab heave, and a case in a drainage adit (Clark & BRE, 2002).

In United Kingdom other case studies were reported by other authors. A foundation of a house garage of 12 years was studied by Hobbs and Taylor (2000). Concrete was manufactured with low Portland cement content, between 120 and 260 kg/m^3. The possible source of sulfates was the groundwater. In the affected concrete thaumasite and ettringite were found. The petrographic examination at the concrete/clay interface have shown a zone where transformation of cement paste to thaumasite was complete. It was also observed a zone of intact concrete with thaumasite and ettringite within voids and a zone with no visual evidence of attack but with ettringite within voids.

The rapid deterioration caused by seawater attack on the bedding mortar of recently constructed harbour wall steps in South Wales was investigated by Sibbick et al. (2003). Within 2 years, the mortar had suffered severe cracking and spalling. The steps were constructed during the spring of 1995. The step construction consisted of a concrete foundation with a reinforced concrete rear wall. Local stonework was then used to face the structure. The stones were bedded and pointed using a 1:1:4, ordinary Portland cement (OPC):lime:sand mix. No records were available of the water/cement ratio. However, during April 1996 (only 1 year after the steps had been built), WS Atkins was contacted by the National Park to inspect some cracking in the masonry after concern had been expressed over the structural condition of the steps and the safety of the public. The initial observation was that the majority of the mortar within the inter-tidal zone had suffered from cracking. On closer inspection, the mortar was found to have a hard crust, which hide a sandy mortar without any cement binder remaining. Once dry, the mortar just crumbled into sand. Believing that the cement had been washed out by the tide, it was recommended that the masonry was re-pointed using a pre-bagged cementitious mortar. These defective joints were repaired during May 1996. Later it deteriorated again. It was observed that thaumasite sulfate attack and extensive thaumasite formation were associated with other deleterious processes related to the attack of normal sea water of cementitious materials, which mainly involved the decalcification of cement paste hydrates. The source of the sulfates clearly appeared to be the seawater, which was replenished by fresh' open seawater with each tide.

A case in Italy of deterioration of a tunnel lining was reported by Baronio and Berra (1984). The concrete lining was in direct contact with the surrounding ground rock and the purpose of the tunnel was to carry a smaller diameter metal pipeline full of water from a dam in the Novara province of Italy to the local power station. The pipeline and the surrounding tunnel had been laid in the 1930s and in 1986, advanced deterioration was noticed in the concrete, with pieces as thick as the whole

300 mm lining breaking off. The deterioration seen was most probably a result of thaumasite sulfate attack and it was so severe that large areas had been transformed into a plastic mass. It was found that pyrite in the rock surrounding the tunnel lining had oxidised forming water-soluble sulfates. These sulfates were then transported to the concrete lining by groundwater, which also contained large quantities of dissolved carbonate ions. The surrounding country rock was used as the concreting aggregate and it contained calcareous schists, another possible source of carbonate ions. The conclusion was that thaumasite had formed due to the low temperatures, very high humidity and presence of CO_2^{3-} and SO_2^{4-} ions dissolved in the groundwater. The absence of ettringite in the deteriorated concrete was due to it having been completely transformed to thaumasite (Baronio & Berra, 1984; Crammond, 2002a, b).

In 1987, in Italy, Berra and Baronio (1987) reported the occurrence of thaumasite in a 40-year old concrete lining affected by compounds resulting from oxidation of pyrites in the surrounding rock. The groundwater was not particularly aggressive—about 260 mg SO_4/l, 10 mg Mg/l, 63 mg CO_2/l and pH 6.5. It is believed that the excavated rock was used as the aggregate, so oxidation of pyrites within the aggregate could have been a major contributor to the thaumasite attack. Tests on the sound concrete gave an MgO content of 1.82% by mass of concrete indicating that dolomite was present. The concrete lining was badly damaged over wide areas, with the concrete reduced to a white plastic mass with pieces as thick as the whole concrete lining (300 mm). In the affected regions, tests showed that thaumasite and calcite were present and that ettringite was absent. Since no comment is made regarding the presence of portlandite it can possibly be assumed that it was absent. The SO_4 content of the plastic concrete was determined to be about 3.3% by mass, namely about 20–25% by mass of the cement (Berra & Baronio, 1987; Hobbs, 2003).

Also in Italy, another case of field structure damaged by sulfate attack was the tunnel lining of Formazzo Valley, dated 1987. Potential sources of sulfates were sulfide-bearing and carbon-bearing aggregates, no direct source of carbonate indicated. The degradation affected the thickness of the tunnel lining. The entire 300-mm thickness of the tunnel lining became detached (Shi et al., 2012).

In France it was identified sulfate attack in the jointing mortar of an old sandstone block-built railway tunnel where gypsum, ettringite and thaumasite were identified as the products responsible for expansion (Deloye et al., 1988).

In Denmark several cases of sulfate attack in field structures were investigated. In 1997 the waterworks in Marbjerg, outside Copenhagen, run by 'Copenhagen Energy', were taken out of production in a short period for inspection purposes. Below the concrete oxidation stairs, in a space not accessible for inspection during the drinking water production, the concrete showed serious signs of surface attack. The attack was found on top of columns and on the underside of the concrete stairs, above the water table. Below the water table no attack was visible. The oxidation of the groundwater is the first step in the series of treatments in the production of drinking water. Groundwater from borings is pumped into a circular concrete tank, 11 m in diameter and 4 m in height. Twelve concrete columns support the superstructure with three concrete oxidation stairs of varying diameter. Through a steel pipe in the middle of the tank, the water is pumped to the top level of the stairs and then runs freely down the stairs, continuing to two filtration units. A small amount of the oxidised water returns into the tank through small pressure equalising windows above the water level. No information concerning concrete composition or possible earlier inspections or repairs was available. The composition of the groundwater, analysed in 1997 and not necessarily representative of the past 65 years, showed no harmful content of aggressive components. An odour of hydrogen sulfide was noticed. The content of sulfide in the water was unquantifiable (limit of detection 0.02 mg/l). The degree of hardness was 21°dH (German degrees of hardness), no aggressive CO_2, pH 7.04, and a sulfate (SO_4) content of 96 mg/l. During the inspection of the empty water tank, columns and the underside of the oxidation stairs the following was observed: Extensive spalling/exfoliation from 5–6 to 10–12 mm depth below the original surface was noted on most concrete surfaces in the space above the normal water table, including top of columns. The surface material was soft and greasy, whitish to brownish, and could easily be peeled off. The material contained sand and small aggregate particles with no adhesion to the binder. The material gave a colourless reaction to phenolphthalein. Beneath the loose surface material the concrete surface, with exposed coarse aggregate particles, was hard and had a sound appearance. The concrete gave an immediate red colour with phenolphthalein after removal of the loose surface material. No signs of reinforcement corrosion were noted, the reinforcement still being embedded in undamaged concrete. Spalling was not observed at places where water flowed back into the concrete tank, seeping down the walls from small pressure equalising windows. The concrete surface not showing attack was covered by a thin layer of red ochre from the groundwater, e.g. on top of the oxidation stairs and on surfaces continuously exposed to water. The surface attack was of a uniform appearance in the air-filled space below the oxidation stairs, e.g. in all places above the normal water level except where water was seeping down the walls. The concrete core without visible signs of attack showed a surface with a thin layer of calcium carbonate and ochre on the surface, and a leached paste of 1–3 mm thickness behind the 0.5–1 mm carbonated surface layer. A few air voids near the surface were filled with calcium carbonate crystals, indicating a weak carbonic acid attack. The core from the column with surface attack not contained the original surface layer. An estimated 7–8 mm of the outer surface was missing. An irregular, weak, and whitish/brownish surface layer was seen in the outermost 6–8 mm of the core, with several surface parallel cracks through paste and aggregate. From the macro observations as well as the microscopic investigations presence of thaumasite as part of the attack were considered most likely under the prevailing conditions: cold, humid surroundings, presence of sulfates (or sulfide), sulfuric acid by help of aerobic bacterial action. A similar attack was typical for sewer systems, just above the water level. In Denmark, thaumasite had not earlier been identified as a part of similar concrete attacks. However, similar cases of damage had earlier been observed in other waterworks, according to Copenhagen Energy'. The repair method had been to remove all loose material and to spray a new mortar layer onto the cleaned surface. As this specific part of the waterworks was not easily accessible no follow-up inspections had been performed.

As the Marbjerg structure was 65 years old at the time of investigation it is concluded that the rate of attack had been low, 0.2 mm/year. The exposure conditions were assumed to have been rather steady with time at the investigated site. A repair solution with a new sacrificial mortar layer of low w/c ratio was decided for the next 65 years combined with regular inspections. At other waterworks conditions could be different, worse or better depending on water composition and presence of hydrogen sulfide (Eriksen, 2003; Clark & BRE, 2000a, b).

Also in Denmark, a diver inspection in 2011 revealed that the concrete below sea level in the Farø Bridges had partly eroded from the surface to depths up to 20–30 mm after about 30 years in service. A special investigation of the condition of the concrete below sea level was initiated in 2012, by use of thin section microscopy of surface concrete to detect the cause of degradation. In this case story diver inspections have shown that for concrete with a sulfate resistant Portland cement and 11% fly ash, at a w/c-ratio of 0.40, sea water attack had taken place to depths up to 20–30 mm after 30 years in service. Petrographic analysis had revealed that sea water attack has progressed to 10–15 mm from the eroded surface from underwater cores, the deterioration primarily caused by sulfate attack (Eriksen & Buhr, 2013).

The cause of degradation of the non-shrinking cement mortar used in repair works in some Slovenian railway tunnels was studied. This cement mortar was used to fill in the joints between the individual worked limestone blocks of the tunnel walls. The results of preliminary investigations showed that the degradation of the mortar was due to the formation of thaumasite. The carbonate ions came from the limestone cladding and from the water, whereas it was assumed that the pollution of the limestone cladding with soot from the furnaces of former steam locomotives, as well as from the fumes of modern diesel engines, was the main source of the sulfate ions, which produced the thin layer of gypsum on the limestone. Understandably, the limestone cladding had not been cleaned sufficiently to remove the old layers of soot and gypsum. Five railway tunnels, located in two different parts of Slovenia, were inspected, and their materials sampled. Two were located in south-eastern Slovenia (Semič, Peščenik), whereas three were in south-western Slovenia (Jurgovski, Ležeški and Križiški). Both of these geographical regions have similar geomorphological and geological characteristics—they are located on the Karst in hard, dense, Cretaceous limestones. However, with respect to their engineering geology characteristics, these limestones could be classified as permeable and liable to the formation of caverns. This it meant that in these areas it was necessary to take into account potential water currents and trickling, depending on the time of year and the quantity of precipitation. The water which penetrated through the ground contained a lot of Ca^{2+} and Mg^{2+} ions, which could participate in the sulfate reactions. The water also contained dissolved CO_2, which was hydrolized into a weak acid and accelerates the dissolving of calcite from the limestone claddings. In the first two tunnels, Semič and Peščenik, the air temperature was between 2.5 °C and 3.5 °C, and the relative humidity was greater than 90%. Water occurred randomly at various locations of the walls and roof of the tunnels in the form of percolation, dripping and bedewing. The air temperature in the other three tunnels (Križiški, Jurgovski, and Ležeški) was in the range between 3.5 °C and 5 °C, with relative humidity greater than 80%. The walls of the investigated railway tunnels, which were more than 150 years old, were heavily polluted. On the surface of the limestone blocks there was a layer of gypsum, on top of which there was layers of soot, iron particles and other smut. Before the repair works were carried out, the surface of the tunnel walls was cleaned using water, but the quality of this cleaning process is open to doubt since, due to the difficult operating conditions (time and space limitations due to railway traffic) only the surface pollution was removed, whereas the layer of gypsum remained on the stones of the claddings. From the available technical documentation it followed that the mortar used to repair the tunnels consisted of one part, by weight, of binder, and three parts, by weight, of quartz aggregate of size 0–3 mm, with small quantities of special additives. The binder was based on Portland cement, but it contained a shrinkage-compensating component, with the chemical composition $3CaO\ 3Al_2O_3\ CaSO_4$. This component amounted to 10% by weight of the Portland cement. On the walls of the tunnels various types of damage were observed such as the swelling and flaking away of the mortar from the surfaces of the stone and the mortar joints. It was clear that the reaction had taken place on the inner side, at the jointing areas between the stone cladding and repair mortar or mortar joints and repair mortar. At these locations the repair mortar, which in its hardened state was usually of a dark grey colour, had changed into a soft, white mushy powder. At those locations where there was greater dampness, a light-grey mud, often in kidney-shaped clumps, could be observed. Among the reaction products of the decomposed, failed repair mortar, thaumasite predominated, followed by gypsum and, mainly in the pores of the mortar, some ettringite. The formation of ettringite and thaumasite in the investigated Slovenian railway tunnels was caused, not only by the general climatic conditions, but also by the composition of the repair mortar (Suput et al., 2003).

The formation of thaumasite in historic mortars was found to be a recurrent problem in cases of conservation of historic masonry in the Netherlands. In several case studies carried out in the Netherlands, in the framework of the EU project "Maintenance of pointing in historic buildings: decay and replacement (contract ENV4-CT98-706)", mortar swelling appeared to be the major cause of the damage. Several cases of swelling salts in mortars, both bedding mortars and pointing mortars were encountered and several types of swelling salts were found. The two cases described here, are concerned with the formation of thaumasite in and swelling of the bedding mortar. The first case is a historic canal bridge (1728) in Amsterdam. The bridge was built with clay brick (a good quality in the external 40 cm and a low strength orange–red clay brick, most probably rich in sulfate, in the internal part). The bedding mortar was a hydraulic mortar (lime–pozzolan binder). The damage mainly occurred in the vaults, where push out of the (cement based) re-pointing and chipping of bricks occurred. The second case is a church tower (1926) in Noordwijk, built in brick masonry, with an external layer being a high strength clay brick and the internal layer a low strength clay brick, rich in sulfate. Originally no pointing existed, but a tooled bedding mortar. Part of the mortar joints have been re-pointed with a cement based mortar. The original bedding mortar was mainly lime based, with some addition of Portland cement. Damage was visible to the wall as a whole:

mainly vertical but also horizontal cracks in the external layer of bricks; vertical cracks through both joints and bricks, horizontal cracks only through horizontal joints; most severe cracking (width up to 17 mm) occurred in the upper part of the tower. In the case of the historic bridge the diagnosis was that the swelling of the bedding mortar was due to the formation of both thaumasite and monochloride. Thaumasite was found up to a depth of 10 cm from the surface. Sulfate may come from inside (the historic brick) and from outside (combustion processes from industry or traffic) (Van Hees et al., 2003).

In Germany several concretes damaged by sulfate attack were investigated. A concrete basin of a saltwater swimming bath and concrete structures from a train-tunnel were investigated. The swimming pool is 68 years old; the concrete formula is unknown. But it contains aggregates from microcrystalline partial biogenic limestone (mittlere Trias, shell-lime). The concrete from the pool-bottom shows no measurable strength. It can be broken by hand. After drying it disintegrates into small pieces. The concrete borders on gypsum and anhydrite containing layers (oberer Zechstein). Also it is in contact with sulfate-containing groundwater. Sometimes the basin was used as a saltwater swimming pool. The saltwater contained sodium- and magnesium sulfate. The tunnel walls built from natural stone were stabilised by concrete structures in 1994. The concrete contained 330 kg/m^3 cement as Portland-limestone cement PKZ 35 F (CEM II/A-L). Under the tunnel bottom there was gypsum- and anhydrite-containing stone. Up to 550 mg/l Ca^{2+} and up to 1670 mg/l Mg^{2+} could be measured in groundwater. After 6 years the concrete structures in contact with the sulfate-containing water was totally destroyed on the surface (up to a depth of 5 cm). In both examples perfect conditions were present for deterioration of concrete by formation of thaumasite: External sulfate attack from groundwater (sulfate from gypsum and anhydrite containing stone), low temperatures (seasonal variation on the bottom of the saltwater swimming pool, constant low in the tunnel) and $CaCO_3$ proportion in concrete (Freyburg & Berninger, 2003).

Also in Germany, in the region of Thuringia several field structures were investigated due to damage by sulfate attack. In this region, due to its geographic history, marine evaporates such as gypsum and rock salt are at or close to the earth crust in some places. The climate is humid with a medium temperature of approximately 8 °C and a medium rainfall of 400–500 mm/year. Bellmann et al. (2012) categorised the structures damaged by sulfate attack following a classification based on exposure environments, associated with dissolved and undissolved sulfur minerals:

- Concrete elevated sulfate concentration in mobile water at normal pH;
- Concrete elevated sulfate concentration in mobile water at reduced pH;
- Concrete in contact with concentrated salt brines (50 to 300 mg salt/kg solution);
- Concrete in contact with gypsum containing soil or rock;
- Concrete in contact with soil and rock containing sulfides.

The characteristics of the structures investigated with concrete in elevated sulfate concentration in mobile water are as follows: Road bridge Kölleda (CEMI) with age 13 years, in which it was observed concrete with low strength, high porosity, not compacted; Road bridge Buttstedt (CEMI) with age 16 years, in which it was observed concrete with high strength, sulfate attack not confirmed; Road bridge Wickerode (CEMI+coal fly ash) with age 15 years, in which it was observed concrete with high porosity, phase transformations up to 1 cm; Road bridge Gera 1 (CEMI+coal fly ash) with age 12 years, in which it was observed concrete with high strength, phase transformations up to 2 cm; Road bridge Gera 2 (CEMIII) with age 35 years, in which it was observed concrete with high strength, bituminous coating, phase transformations up to 1 cm; Concrete basin Elbingerode (CEMIII) with age 15 years, in which it was observed concrete with phase transformations up to 1 cm, deterioration when exposed to frost Outflow channel Dachwig (CEMIII) with age 33 years, in which it was observed concrete with 1 cm already lost, phase transformations in further 2 cm.

The structures investigated with concrete in elevated sulfate concentration in mobile water at reduced pH were the following: Water channel Cospuden (CEMIII+coal fly ash) with age 3 years, in which it was observed concrete with high strength, phase transformations up to 0.5 cm; Bridge Barbarakanal (CEMI) with age 6 years, in which it was observed concrete with high strength, acid attack, phase transformations; Concrete basin Restloch 29 (CEMIII+coal fly ash) with age 2 years, in which it was observed concrete with high strength, acid attack, phase transformations;

The case studies investigated where concrete is in contact with concentrated salt brines were the following: Sewage manhole Menteroda (CEMI SRPC) with age 6 years, in which it was observed concrete with low strength, concrete completely deteriorated; Concrete basin Bischofferode (CEMIII+coal fly ash) with age 7 years, in which it was observed concrete with high strength, phase transformations up to 2 cm, chloride attack.

The structures investigated with concrete in gypsum containing soil were the following: Bridge Rottleberode (CEMIII) with age 47 years, in which it was observed concrete with destruction not due to sulfate attack, phase transformations up to 1 cm; Radio mast Vogelsberg (CEMIII+coal fly ash) with age 10 years, in which it was observed concrete with high strength, phase transformations up to 2 mm; Power transmission line Nordhausen (CEMI) with age 44 years, in which it was observed concrete with low strength and carbonation; Soil stabilization Leinefelde (CEMI+lime) with age 1 years, in which it was observed concrete with Formation of ettringite and thaumasite, up to 15–25 cm expansion.

Structures investigated with concrete in contact with soil and rock containing sulfides were the following: Tunnel Lössau (CEMII, CEMIII) with age 30 years, in which it was observed concrete with CEM II shotcrete disintegrated in some spots, minor problems with CEM III concrete; Tunnel Leibis/Lichte(CEMI) with age 10 years, in which it was observed concrete with shotcrete completely disintegrated in some spots; Tunnel Breitenbachtalsperre (CEMI, CEMIII) with age 42 years, in which it was observed concrete with CEM I shotcrete disintegrated, CEM III shotcrete unaffected; Water channel (CEMI SRPC) with age 2–4 years, in which it was observed concrete with shotcrete completely disintegrated in some spots.

Six out of 20 structures mentioned above show serious damage and demand immediate repair (Sewage manhole Menteroda, Soil stabilization Leinefelde,

Tunnel Lössau, Tunnel Leibis/Lichte, Tunnel Breitenbachtalsperre, Water channel). Most of them showed disintegration due to thaumasite sulfate attack (Soil stabilization Leinefelde, Tunnel Lössau, Tunnel Leibis/Lichte, Tunnel Breitenbachtalsperre, Water channel). Thaumasite formation was identified in structures samples of Road bridge Kölleda, Road bridge Wickerode, Road bridge Gera 1, Outflow channel Dachwig, Sewage manhole Menteroda, and Radio mast Vogelsberg.

Solid materials and aqueous solutions were sampled in Austrian highway and railroad tunnels to assess sulfate attack. Intense alteration at the border of the old brick wall lining (early 1900) and on the surrounding shotcrete that was installed in the 1960s had caused large areas of the shotcrete to fall off, resulting in endangering the traffic that passes through the tunnel. Heavily damaged concrete from the tunnels consisted mainly of thaumasite, secondary calcite, gypsum, and relicts of aggregates. Synthesized ettringite contained small amounts of calcite while synthesized gypsum was pure. The concrete completely loosed its strength and the cement matrix gets destroyed by thaumasite and gypsum formation. Surprisingly, the groundwater and the drainage water analyses generally revealed low to moderate SO_4 contents, with maximum values of about 500 mg L^{-1}. Hence, sulfate attack had mainly appeared in environments with generally low SO_4 concentrations and the underlying reactions were yet to be understood. Evaporation was suspected to have caused the critical level of sulfate concentration, which had finally led to the TSA-induced concrete damage. This highly dynamic system of drying and wetting cycles causes large degradation (Mittermayr et al., 2013; Mittermayr et al., 2015).

In Switzerland, the concrete lining of a vertical ventilation shaft exposed to sulfate-bearing groundwater for 45 years was investigated. The concrete was characterized at several sites with different degrees of damage. The concrete was produced using a low C3A-type Portland cement and aggregates containing calcite. Well-rounded alluvial gravel consisting of limestone, sandstone and igneous rocks with a grain size distribution of 0–32 mm was used as aggregates. The calcite content of the aggregates determined using thermogravimetry analysis was 7% mass. Its cement content might have been 300–330 kg/m^3 and its water-to-cement ratio (w/c) 0.55–0.65. Concrete deterioration led to loosening of the anchors of the elevator and severe spalling poses a constant danger during utilisation of the open elevator. The distribution of damages showed that the groundwater was able to penetrate the concrete and reach the inner surface of the lining only in areas of poor concrete quality or low concrete thickness. In such areas, the concrete showed severe spalling. The sulfate resistant cement used for concrete production was not able to prevent the damages. Where no such weak areas existed, sulfate ingresses mainly by diffusion. Due to the slowness of the process, sulfate penetrated only a few centimetres at the outer surface of the concrete lining that was exposed to the rock face and the groundwater. The main damage on the concrete lining was caused by ettringite induced expansion leading to severe spalling at its inner surface. In this area, thaumasite was mainly formed in cracks created by the ettringite-triggered expansion. Along the rock face, no cracking or spalling occurred as the concrete was restrained. Here the cement paste was transformed *in-situ* to thaumasite and therefore disintegrates nearly completely. Dissolution of calcite and portlandite showed that both are used as a source of carbonate and calcium respectively for thaumasite formation. Gypsum was only present in already severely deteriorated areas very close to the rock face with a SO_3-content above 20% mass (Leemann & Loser, 2011).

Also in Switzerland, several concrete railway tunnels with signs of damage by sulfate attack were studied. In all investigated tunnel structures with concrete in contact with groundwater, the processes of leaching and the formation of sulfate minerals, predominantly thaumasite, had been detected (Romer, 2003).

The hardened cement paste of 30-year-old shotcrete of Bauen-Tunnel in contact with the crack was completely amorphous, representing the corrosion zone. The alteration product in addition was affected by smaller cracks perpendicular and parallel to the larger cracks. Hardened cement paste showed in dark grey which indicated the loss of portlandite. The sample showed extensive cracking parallel to the contact of the shotcrete. Cracks and larger veins were filled with fibrous aggregates of thaumasite. In San Bernardino tunnel (service gallery) in some areas the concrete was completely altered into a mush-like material. Concrete presented mineral aggregates as quartz, feldspar and mica. Hydrated cement paste showed darker grey levels and granular texture. Thaumasite and subordinate ettringite was observed. The nearby groundwater was analysed to contain, in mg/l, calcium 270, magnesium 270, bicarbonate 200 and sulfate 1900. The original concrete was produced with sulfate resistant cement. Some years after construction of the tunnel the drainage system collapsed, the pressure of the groundwater increased and percolating water was forced to penetrate the unprotected concrete lining of the tunnel. For instance in the San Bernardino tunnel in regions with thaumasite-related concrete, deterioration the temperature was always above 20 °C. In the case of tunnel linings consisting only of shotcrete the contact with low amounts of weakly mineralised groundwater was leading to partial break down of the concrete due to the formation of thaumasite on the back of the lining over 20–30 years. Potential source of sulfates were the followings, sulfur-bearing soot layer on the concrete lining, and leaking drainage. Larger pieces of shot-concrete could be detached from their support (Romer, 2003; Shi et al., 2012).

In Switzerland, others structures with damages by sulfate attack were investigated. This concrete structures were as follows: Precast concrete in drainage system in Hauenstein tunnel, located in Olten; sulfate attack and anhydrite swelling in Belchen tunnel; Concrete pipe, Tiefenkastel (EWZ Kraftwerke); Formation of ettringite, thaumasite and crack formation and loss of adhesion by sulfate containing soot of steam trains and Shotcrete on rock and masonry in Tunnel Koblenz; shotcrete in Exploratory tunnel Visp; Shotcrete (CEM I HS) in Gotschna tunnel; shotcrete in Harderband tunnel (Leemann and Loser, TC 251-SRT presentation).

Also, it was found thaumasite in the gunite lining to the Dettenberg rail tunnel in Switzerland (Schwander & Stern, 1988).

Thaumasite formation and limited thaumasite form of sulfate attack had recently been detected in several Norwegian sprayed concretes. A collaborative investigation programme was set up by the Building Research Establishment (BRE), UK and the Norwegian Public Roads Administration (NPRA), Norway in 2000. Several cases of concretes damaged by sulfate attack were investigated. They had found thaumasite formation, TF, and TSA in high quality sprayed concrete in Oslo, even in concretes under 2 years old. Thaumasite formation and thaumasite sulfate attack were frequently associated with simultaneous and late decalcification of both C-S-H and thaumasite, which in both cases led to the formation of unusual forms of internal secondary calcite. The sprayed concretes investigated were the followings (Hagelia et al., 2003):

- Road cut in Äkeberg, Oslo, with 13 and 30 years, in ground conditions of alum shale. This first documented occurrence of thaumasite in Norwegian concrete consisted of 2–3 layers of steel fibre reinforced sprayed concrete made in 1987. The concrete was made with sulfate resistant Portland cement, SRPC, and 5–10% silica fume, and water glass as setting accelerator. The w = c-ratio was about 0.45–0.5. TF and limited TSA occurs quite near the open outer surface immediately beneath ordinary surface carbonation, in vicinity of the concrete–Alum Shale interface, as well as internally in entrapment voids between layers and in their surroundings. Different types of thaumasite were encountered. At present this concrete had become deteriorated with extensive rusty surface deposits from oxidation of sulfides as well as calcite efflorescence being apparent. The aggregate did not contain any sulfate and carbonate source materials, except from minor amounts of calcite fines. The concrete was consistently moist with scattered tiny leakages throughout the spray. Spalling had started to develop in association with a 10–20 mm wide layer of thaumasite +calcite (mush zone). Secondary internal calcite formation frequently involved the decalcification of C-S-H. This was closely associated with calcite ± subordinate gypsum formation at the expense of thaumasite. This breakdown reaction occurred both within the intact sprayed concrete as well as at an internal surface against the thaumasite+ calcite mush zone, which was in contact with air. Presence of gypsum tends to be more common near the zone of ordinary surface carbonation. Calcite deposits with no obvious relationship to thaumasite or C-S-H were also frequently encountered in air- and entrapment voids. In general the calcite varied in form from blocky, branched/dendritic, popcorn- and cornflake like forms to acicular crystals. One water sample taken as it leaked from a crack through the sprayed concrete had a pH of 7.6. This water composition is probably not very different from unaffected groundwater behind. However non-oxidised groundwater's in Alum Shale may be slightly acidic due to predomination of ferrous iron. As regards oxidised water in the ford zone, pH may be 3–4 or slightly less, due to oxidation of ferrous iron and hydrolysis of ferric iron. Thaumasite formation, thaumasite sulfate attack and carbonation had also attacked remnants of a previous sprayed concrete layer from 1970. This concrete was made with sulfate resistant Portland cement but without silica fume.
- Highway tunnel in Ekeberg, Oslo, with 8 years in ground conditions of alum Shale/clay stone. The steel fibre reinforced sprayed concrete in this tunnel was placed in 1992, and the concrete mix was essentially the same as at Åkebergveien. w/c ratio was about 0.45. Due to previous experience sprayed concrete was not applied on wet alum Shale, but instead the rock surface was dried (as much as

possible) using compressed air just before the spraying operation in order to achieve bonding. Silica fume (ca. 8% by cement weight) was in this case well dispersed throughout the matrix. Yet thaumasite formation, and limited thaumasite sulfate attack were found. Calcite stalactites and rust deposits were commonly observed in presence of water leakages throughout the concrete. This was usually also associated with steel fibre corrosion in several places, although apparently not usually very severe. The overall TF-TSA internal carbonation process was somewhat less well developed than at Åkebergveien, but all features were broadly similar. Different types of thaumasite were encountered. The pH range within the groundwater was about 4–5 within this Alum Shale sequence just after blasting. Water samples collected in 2000 from leaks associated with calcite stalactites as well as rusty brown deposits along a crack through the concrete had a pH of about 7. The water compositions may have been slightly modified by the concrete.

- Highway tunnel in Svartdal, Oslo, with 2 years in ground conditions of alum Shale. The spraying method and concrete mix were essentially similar to the Ekeberg tunnel, although it also contained a small amount of limestone aggregate. The reported w = c ratio was slightly above 0.4, and they had found that silica fume was very well dispersed. In fact all cores were extracted the day before the tunnel was officially opened for traffic in August 2000, when the sprayed concrete was already about 2 years old. Surprisingly the sulfate attack process was also found in this younger concrete. The steel fibres were occasionally slightly corroded, when in contact with the secondary calcite. Thaumasite sulfate attack within most samples was restricted to rather thin zones (<1.5 mm) in the vicinity of the concrete Alum Shale interface and elsewhere. Although secondary calcite frequently had formed at the expense of C-S-H and thaumasite, additional euhedral calcite coexisted with thaumasite in some large entrapment voids. Some of them were multiple crystals resembling popcorn calcite. This paragenesis represented simultaneous deposition of thaumasite and calcite, and not partial thaumasite decomposition. They had also found occasional evidence of a later stage partial dissolution of this assemblage, most likely due to the action of somewhat acidic water.
- Civil defence shelter in Oslo, between 25 and 35 years in ground conditions of alum Shale (swelling). Sprayed concrete (dry method, fibre free) was applied directly on Alum Shale in the late 60s or early 70s. Due to the earlier focus on sulfate attack in the Oslo region it was expected that all concrete constructions here were made with SRPC. This was however not easy to confirm due to severe deterioration. In 2000 about 1–5 m^2 areas of very friable sprayed concrete had spalled and fallen down in several places. Jarosite and gypsum were frequently observed at the now exposed Alum Shale, and internal swelling within the outer surface of the shale had obviously contributed a lot to this extreme spalling. Rust deposits occurred on the surface. In the sprayed concrete thaumasite formation and thaumasite sulfate attack were clearly present as a major cause of the observed degradation, at least within a 5 mm thick layer in contact with the Alum Shale. The concrete adjacent to the shale was usually represented by microcracks

heavily filled with secondary carbonate, some of which had apparently been derived from thaumasite. Also abundant gypsum crystals as void filling, plus ettringite were observed to be contributing to degradation. The main degradation was still in the form of abundant surface parallel microcracks variously filled with thaumasite, gypsum, ettringite and calcite, which were co-habiting within relatively small domains. These sulfate deposits were often closely associated with abundant calcite deposits probably secondary, which were filling pre-existing surface parallel microcracks within the cement paste and aggregates particles. Mottled and popcorn-like carbonation was also observed within the cement paste in several places, and must have contributed to the overall loss of strength.

- Access tunnel to hydropower plant in East Norway, with 16 years in ground conditions of metasediments with black gneiss. The ground conditions here were characterised by somewhat elevated sulfate and calcium contents in groundwater, as evidenced with historically known surface precipitation of bitter salt (gypsum with some epsomite), due to the rather dry climate in this part of Norway. The bedrock was characterised by various quartz–feldspar–mica gneisses and black gneiss ("shale"), which is equivalent to Alum Shale having additionally suffered higher temperature deformation during the Caledonian Orogeny. Locally within the tunnel jarosite and some gypsum had formed by weathering of the black gneiss. The 1984 fibre free sprayed concrete was made with Portland Cement with about 8% C3A and most likely with Fly ash. Water glass was used as setting accelerator. The locals said that the tunnel was initially characterized by abundant water leaks, which was the main motivation for a sprayed concrete coating. In fact, extensive white surface efflorescence was present, thus confirming the presence of significant aqueous action at an earlier time. There was also several oxidized brown deposits on both the surface of the concrete and the rust-colored muds at the base.
- Sub-sea road tunnel in Freifjord with 10 years, in ground conditions of modified sea water. This tunnel was opened for traffic in 1992, and steel fibre reinforced sprayed concrete was made with about 6–8% silica fume and Norcem RP 38: a rapidly setting Portland cement that contained about 8% C3A. Water glass was used as a setting accelerator and the reported w = c ratio was in the order of 0.37–0.47. Systematic investigations including tunnel mapping and laboratory analysis were undertaken in 1996 as part of a major project on durability of sprayed concrete. It was concluded that there was no evidence of 'chemical attack on the local concrete lining. However, by 1996 calcite stalactites were abundant and in several restricted areas the concrete had been partially to completely transformed into a mush. These deteriorated concretes were located within a sequence of banded gneiss with layers of marble, and were typically associated with water leakages. The mush occurred as loose layers on the surface of more sound sprayed concrete and had usually also been accumulated as a sludge at the base. This phenomenon was thought by some authorities to have been caused by rebound during the spraying operation, while others claim this was not so. Investigations based on about 20–100 mm size samples of less dete-

riorated white and friable remnants of sprayed concrete, within the now washed away mush, clearly shows that deterioration was caused by thaumasite sulfate attack and magnesium attack. Both thaumasite and brucite had partly replaced C-S-H. Even in fairly solid remnants of an initially 50–100 mm thick spray, full scale thaumasite sulfate attack was observed. Thaumasite formation had occasionally formed along microcracks that post dated the paste matrix, which had already been transformed into a friable thaumasite sulfate attack depleted C-S-H assemblage. Thaumasites were found. Secondary carbonation was very extensive in all forms described above, and some gypsum and ettringite were also detected. The steel fibre reinforcement was frequently completely transformed into a rusty Fe-hydroxide most notably in presence of abundant popcorn calcite and other forms of calcite. Published water analyses from the thaumasite sulfate attack affected tunnel sections, probably collected from cracks in vicinity of degraded concrete, had pH from 7.0 to 7.3, high contents of SO_2^{4-}, Mg^{2+}, Cl^- and very high Ca^{2+}, suggesting modified seawater. Calcite in the marble layers could represent a potential carbonate source for thaumasite and secondary internal calcite. However, calcite was very slightly soluble at pH about 7 and calls upon another carbonate source. Hence a contribution from atmospheric CO_2 should not be excluded.

Also in Norway were investigated others structures damaged by sulfate attack. Concrete basement beneath the animal rooms were investigated (Manure cellars of closed connection). Mainly structure elements above are manure level close to trap doors. The dates of construction were 1931–1983. Excess salt in the urine was soaking the concrete, probably with some drying and wetting cycles. And the carbonate source for the thaumasite formation could also be urine (Cattle was licking salt blocks and excess chloride was excreted into urine). Carbonate source for thaumasite could either be carbonation by enriched CO_2 from fermenting manure or reaction between urea and calcium hydroxide. Manure cellars represented very aggressive environment to concrete with CO_2 and H_2S gases and liquids with urea, chloride, sulfates, nitrates and ammonium. Deterioration of concrete and reinforcement corrosion in the manure storage cellars were observed. Gypsum, ettringite and thaumasite were presented. Concrete bridge of 10 years was studied. Lower edge of the I-beam of a bridge was damaged. Seawater and Seagull guano were the principally sources of sulfate. Strong chemical alterations in the first 5 mm from the surface were the type of degradation observed with thaumasite and ettringite. Also pipes of asbestos reinforced cement after 30 years in service were investigated. Seawater was the principal source of sulfates. Cracked open in pipe walls. The wall thickness was 36 mm and cracks along the surface to a depth of 8 mm from the inner. In addition, there was a 1.5 mm layer of another density/color at the inner surface. Thaumasite was observed (Hustnes & Rodum, 2006).

Also three 12-year-old wall elements had been stored in the tidal zone located in Norway, the Trondheim fjord. All elements were made of ordinary Portland cement, filled with limestone, fine sand 0–2 mm, sand 0–8 mm, crushed gravel 8–11 mm, crushed gravel 11–16, water added, 10% Of MicroAir at 5 ± 1% by volume air,

Scanflux CP30 plasticizer, Na-Gluconate retarder, 33 to 50% calcium nitrite. Close-up of the skin spalling of wall element was the type of degradation. Ettringite and thaumasite were observed (Ostnor et al., 2014).

In Spain several cases of concrete structures with damaged by sulfate attack were investigated. These structures were the followings: Concrete with external sulfate attack in Mequinenza Dam, Saragossa; Concrete with internal sulfate attack in Paso Nuevo Dam, Huesca; Masonry mortar used for the reception of tiles, and also also used in the pavement, of a rural construction in the municipality of Los Yébenes (Toledo); Mortar used to repair a fissured area in the Rumedo dam, located on the river of the same name (Ebro slope) near the town of Lladorre (Lérida); Concrete surface of the embankment in the stirrups of the Viaduct of Pallaresos, in Lérida, after having been treated with jet grouting; Base made in soil-cement that underwent expansions in the highway C154 Muntanyola—Oristà, in Barcelona. In the various investigations, it was most likely that the main source of sulfates are the aggregates used in different concretes (Chinchón, 2013).

Also in Portugal it was observed a concrete structure with damage by sulfate attack. In dock at harbor of Leixões, manufactured in 1937–1938, mortar expansion was observed due to reaction between sulfate from seawater with reactive alumina from a granitic aggregates having partially kaolinized feldspars. The source of sulfates was the seawater.

3.2.3 Asia

China has a vast territory, and in many regions conditions are conducive to thaumasite formation in concrete. Many hydraulic concrete structures, such as Yongan Dam, have been reported to be deeply deteriorated for over 6 years. In August 2004, the investigations on the concrete of Yongan Dam were made. Yongan Dam is in Keshi, Xinjiang where the land is arid and rich in various kinds of salt especially sulfate salts. Furthermore, this area is characterized by large daily thermal amplitudes that could reach more than 20 °C. The construction of the dam was finished in August–October 2003; stored water in November 2003; and dried again in winter. In April 2004, after water storage in March, the slabs had become grey and mushy in the areas where contact with groundwater occurs. No visual signs of degradation have been observed in the areas that do not contact with groundwater. The analysis of the cause of concrete deterioration was associated to a combination of factors: on the one hand the geographical conditions of the Yongan Dam, namely the low temperature at most times of the year, and on the other hand the raw materials used to produce concrete. Actually, concrete with 300 kg/m^3 of cement and water/cement ratio 0.5 was produced using calcareous aggregates, a cement containing about 9% of limestone and mixing water with high sulfate concentration of, 653 mg/l SO_4. All these conditions favoured thaumasite formation in concrete (Hu et al., 2004, 2006).

A site investigation and sampling was carried out on a sulfate-attacked concrete structure in Bapanxia Hydraulic Power Plant in Western China. Bapanxia Hydraulic

Power Plant is located on the Yellow River in Xigu District of Lanzhou City, China and was constructed between 1969 and 1975. The irrigated plateau was situated on the north bank of the dam along the river, where the soil and the groundwater contain a high concentration of sulfate. Observations within the tunnel suggested that the run-off water, which was rich in sulfate, percolated from the outer surface through cracks and preformed ducts to reach the tunnel, where it collected in the drainage channel. At the points of entry, the tunnel wall concrete was also seriously attacked, particularly at the northern end which was closer to the irrigated plateau. In some areas, deep deposits of white crystalline salts were also seen deposited on the tunnel walls, confirming the flow of mineralized water through the concrete into the tunnel. The northern side tunnel was the most severely damaged by sulfate attack and had been repaired every few years since the dam was constructed. The last restoration was in 1997, when all the damaged surface concrete was replaced with a high-quality cement concrete made with ordinary Portland cement and a natural gravel coarse aggregate with a maximum size of 10 mm. In 2004, most of the floor and wall concrete appeared to be undamaged, but there was very severe deterioration along edges of the drainage channels at either side. Here, the concrete shows expansion, cracking and softening in some regions with the maximum depth of damage being more than 100 mm. Some of the concrete lost its strength completely and could be easily crumbled by hand, in which case the cement paste transformed into a white, pulpy mass with no strength. The classic form of sulfate attack and carbonation of concrete appear to have occurred and eventually the internal surface concrete of the tunnel has been decomposed into a white, non-binder, pulpy material, consisting of a mass of thaumasite crystals, together with crystals of ettringite, gypsum, calcite (Ma et al., 2006).

Also the sprayed concrete linings of Chinese railway tunnels located in Chuxiong city of Yunnan Province were seriously damaged by the sulfate attack. Fala tunnel and Bijiguan tunnel, located in Chuxiong city of Yunnan Province were inspected. And the concrete samples cored. These tunnels were built about at the beginning of 1965. This region locates in the south-western region of Yun–Gui Altiplano, belonging to sub-torrid zone, and the year-average temperature is about 12 °C. Secondary sulfate minerals such as gypsum formation under very high concentration sulfate ion condition by accumulating and evaporation process, followed by the crystallization of sulfate salt and formation of thaumasite were found. The deterioration of concrete was accelerated by factors such as poor quality of concrete, shotcrete linings thin and inhomogeneous, low strength and high permeability concrete and the low relative humidity caused by frequent running through of train speedups the transport process and even accumulative or crystallization role of aggressive salt solution of concrete. With respect to engineering geological characteristics, the main rock includes the red sandstone, mud-based sandstone and gypsum mine in cretaceous. The red sandstone was also used to build the side-wall of tunnel. The Fala tunnel investigated has a length of 2100 m. The vault (roof) of tunnel was supported by shotcrete and the side-wall was built by red sandstone and mortar or concrete. Serious damage of the whole tunnel including tunnel inlet and inside was observed though three larger scale repairs had been made in 1982, 1987 and 2000, respectively. And small scale repairs and maintenances are almost carried out each year according to the related records (Long et al., 2011).

The railway tunnel concrete lining exposed to sulfate-bearing groundwater beyond 40 years in southwest of China, Cheng-Kun railway line, was investigated. Field investigation, sulfate ions content and corroded products analysis, macro/microscopic test were carried out. Results showed that under the tunnel concrete lining structure and its served environmental conditions, sulfate solutions permeated concrete lining and accumulated on windward-side of concrete lining, resulting in the increase of sulfate ions content on windward-side and the diffusion of sulfate ions from windward-side to waterward-side, which caused the concrete lining of windward-side damaged seriously but the waterward-side of concrete lining was still in perfect condition. It was confirmed that structural characteristic of tunnel and environmental conditions leaded to physical attack with the leaching of concrete and sodium sulfate crystallization as well as chemical corrosion with formation of gypsum in high sulfate concentration and formation of thaumasite in proper temperature rather than formation of ettringite (Ma et al., 2012).

More than 10 existing tunnels lining in the Chengdu–Kunming railway were investigated. The results showed that the salt medium usually enters the tunnel concrete lining by means of capillary absorption, permeation, and diffusion. In turn, these lead to liner concrete degradation and spalling, especially when combined with the effects of temperature and humidity. The damage styles included serious corrosion of the concrete lining, crystallization erosion, and spalling on the lining surface. Along with serious deterioration occurring in the liner of the leakage areas, the formation of loose corrosive matter without bond was observed. In investigations, deterioration occurred up to a depth of 60 mm from the liner surface. Meanwhile, serious crystallization corrosion was found on the surface of the tunnel lining from the side wall to the vault. The salt content was extremely high both on the lining surface and in the lining itself; and these salts easily led to crystallization. Another main type of tunnel lining damage due to sulfate ambient was surface spalling, which was observed on the side wall and vault. The quality of the concrete lining was poor, characterized by low strength and high permeability, the erosive medium can diffuse into concrete inner easily; this is especially true in the old tunnels. According to the monitoring data supplied by the railway division of this line and the water analysis results, the stratum located by existing tunnels in the Chengdu–Kunming railway was rich in undergroundwater with great amounts of SO_4. Meanwhile, field investigation showed that the concrete surface of the tunnel lining was covered with a thick layer of black pollutant, including smoke dust produced by locomotive (especially the diesel in early stage) and sulfur-containing substances from coal transportation during tunnel operation. The pollutants produce aggressive sulfate and the medium permeated into the inner concrete lining of the tunnel from outside. Moreover, if the leakage points are not sealed effectively during the tunnel construction, the tunnel concrete is directly lixiviated by the aqueous solution with sulfate, resulting in greater concrete damage (Lei et al., 2013).

In Thailand the reinforcement concrete structures in Kasetsart University, Kamphaengsan campus, were investigated. The environment characterized by high humidity and temperature and excessive soil borne salinity has resulted in corrosion of the reinforcement and subsequent cracking and spalling of concrete. The purpose of soil salt investigation was to determine the cause of deterioration on concrete structures. The amount of SO_4 in soil was significantly high between all chemical compositions and Na was the second high content. The results of sulfate demonstrated that the concentration of sulfate can be the cause of damage in concrete structure around Kasetsart University, Kamphaengsan campus (Khomwan et al., 2009).

3.2.4 Africa

In 1983, was detected thaumasite in concrete blocks containing a black carbonaceous shale within which there was a high content of disseminated fine pyrrhotite, pyrite and chalcopyrite giving a sulfur content of 1.44% by mass and a potential SO_4 of 4.3% by mass. Extensive random cracking was apparent in the rendering of external walls. Wide vertical cracks in the lower portions of the external walls were attributed to expansion of the floor slabs, which also contained black shale as aggregate. Thaumasite was observed around the aggregate particles and throughout the mortar. The authors attributed the thaumasite attack to the oxidation of pyrites with the sulfuric acid formed reacting with calcite in the aggregate and the cement hydration products to form gypsum (Oberholster et al., 1983; Hobbs, 2003).

In 1984 severe expansion and cracking, with white, powdery deterioration products were observed in concrete brickwork of several houses of Eastern Transvaal, South Africa. The Possible influence factors were Carbonaceous, sulfide-bearing cummingtonite slate aggregates, no mobile water, no direct source of carbonate, ambient temperatures up 15 °C (Shi et al., 2012).

3.3 Analysis of Damaged Structures: Identification of the Most Common Situations

Deterioration or damage of concrete structures can be driven by different conditions. To analyse the process a classification scheme was devised.

The conditions considered for analysis are classified according to the following groups:

- According to the environment in which is located the concrete structures:
  - Source of sulfates (from soil, from water or groundwater, from the air, seawater, aggregate, others)

  - Origin of sulfates (gypsum and anhydrite, oxidation of sulfides, alkali sulfates, other origin)
  - Type of environment (constantly wet, often wet, with wetting/drying cycles, often dry)
  - Temperature (cold, mild, hot)

- According to concrete elements (foundations, dams, tunnels, others like pipes, steps, houses, waterworks, pavements, etc.)
- According to the materials:
  - Type of binder (SR cement, common cement, others)
  - w/c of the concrete mix (less than 0.4, between 0.4 to 0.6, more than 0.6)
  - Binder content (less than 260 kg/m^3, between 260 and 320 kg/m^3, more than 320 kg/m^3)
  - Type of aggregates (calcareous, siliceous, both)
  - Porosity and permeability of concrete (low, medium, high)

- According to degradations:
  - Time of appearance of the first signs of degradation (less than a year, a few years, a few decades)
  - Speed of degradation afterwards (slow, medium, fast)
  - Type of degradation (cracking, spalling, swelling)
  - Degradation depth (superficial less than 1 mm deep, several cm deep, general)
  - Mineral-related degradation (ettringite, gypsum, ettringite+gypsum, others)

- According to impact of degradations on the integrity of the structure (weak, medium, strong)

Analysis of the damaged structures according to the above classification allowed to identify the most common conditions:

- According to the environment in which is located the concrete structures, the most common sources of sulfates are from soils, from water and groundwater. The main origin of sulfates is gypsum and anhydrite, and oxidation of sulfides (generating sulfate ions + low pH). The type of environment more frequently found is with wetting/drying cycles. Most usual temperatures are cold.
- The types of structures most attacked are foundations and tunnels
- According to materials, the type of binder most common is with common cement. w/c of the concrete mix is between 0.4 to 0.6. Binder content is more than 320 kg/m^3. The type of aggregates more common is calcareous. The porosity and permeability of concrete is high.
- According to degradations, the time of appearance of the first signs of degradation is a few years. The speed of degradation afterwards is generally fast. The type of degradation is cracking, spalling and swelling. The degradation depth is several centimetres deep. And the crystallizations or mineral-related with the degradation are, ettringite, thaumasite and gypsum.
- In general the impact of degradations on the integrity of the structure is weak.

3.4 Parameters Influencing Sulfate Attack on Concrete Structures in Service

3.4.1 Water/Binder Ratio

It is well known that cement hydration products are balanced with pore solution. The ingress of external ions changes this equilibrium, so the dissolution of some hydration products occurs to form new products (Aïtcin, 2004). As in other deleterious processes, transport mechanisms by which sulfate ions ingress into concrete are determinant for the deterioration rate. Therefore, the reduction in the permeability of concrete is a major controlling tool regarding sulfate attack (Mehta et al., 1992), and this is mostly achieved by reducing the w/b ratio.

General recommendation is that the higher the aggressiveness of the environment, the lower must be the free water/binder ratio of concrete. According to BRE Special Digest 1 (BRE, 2001), it should vary from w/c = 0.35 for high aggressiveness to w/c = 0.55 for low aggressiveness. In this sense, Khatri and Sirivivatnanon (1997) related expansions recorded in concretes with different strength levels and cement types with their permeability. Here, the type of cement showed a secondary role, intervening in the formation of expansive products but with little relevance for the penetration rate. This study included concretes made with ordinary Portland cement (OPC), high slag content cement (HSC), sulfate-resistant cement (SRC), OPC + 7% silica fume, and HSC + 7% silica fume, with the specimens immersed in 5% Na_2SO_4 solution. Similar conclusions were found by Brueckner et al. (2012), who indicate that deterioration rate by TSA is directly dependent of the w/c ratio and cement content.

Similarly, Brown and Hooton (2002) found that an increment in the water/cement ratio of 0.05 (from 0.45 to 0.50) significantly increased the damage produced by sulfate attack in concretes immersed in sodium sulfate as well as in magnesium sulfate.

As well, data from structures in service showed that the use of a low water/binder ratio in repair mortar in railway tunnels in Slovenia was effective for reducing the extent of the degradation due to thaumasite in presence of favorable environmental and pollution conditions for its formation (Šuput *et al*, 2003).

3.4.2 Type of Aggressive Agent

The type of aggressive agent may vary the deterioration rate in a very significant degree. This is not only depending on the associated cation, but also on the mineralogical form which would allow more or less mobility and reactivity of sulfur species.

Moum and Rosenqvist (1959) described the aggressiveness of metamorphosed shales containing an unstable iron sulfide, pyrrhotite. In the unmetamorphosed shales of central Sweden the sulfur occurs as pyrite (FeS_2), but in the Oslo area part of the sulfur is present as monoclinic pyrrhotite with composition $FeS_{1.14}$. The content of this mineral was directly connected with the aggressiveness of shale, which detrimental action can be exceptionally severe and very fast (with extreme examples of underground concrete walls transformed into mush in only 9 months). A field study performed in the Blindtarmen tunnel with these exposure conditions included specimens permanently submerged and specimens located in between the variable water level. After 3 to 4 years the prisms the low-water level still remained with sharp edges and fairly good mechanical performance, whereas those submerged were mostly destroyed, excepting those made with sulfate resistant cement. Large amounts of ettringite were determined in the damaged specimens, but no gypsum or anhydrite was found. The mechanism of attack was connected with the composition of sulfate solution that contains bivalent iron, which is later more or less oxidized to trivalent iron. In the samples from the upper part of the tunnel the conditions were different. The water there was acid and contained trivalent iron. Eventually this trivalent iron will be precipitated upon the surface of the concrete, fill the pores and make the concrete more watertight, so that further penetration is slower. Sodium sulfate solution penetrates into the concrete and when sulfate reacts, sodium hydroxide will increase the pH. In contrast, bivalent iron sulfate may penetrate the sound concrete in the same way as the sodium sulfate solutions and carry with it higher concentrations of sulfate. In the interior of the concrete it will precipitate insoluble iron hydroxide which may oxidize, and there will probably be a decrease in pH. Here, a combination of sulfate and acid attack takes place.

The formation of thaumasite requires the presence of carbonates, which are generally provided by limestone cement or limestone aggregates. However, dissolved CO_2 and carbonates in the water in contact with concrete being a source of CO_3^{2-} might also promote the formation of thaumasite. Also the process of dedolomitization in dolomitic aggregates can provide sufficient carbonates for thaumasite formation (Thomas et al., 2003). Even the oxidation of carbon in the form of graphite contained in some types of rock may be a good source of carbonate (Loudon, 2003). Graphite can oxidize in contact with acidic solution originated from the decay of sulfides present in quartzite and slate rocks. This process would explain the degradation by TSA observed in a concrete tunnel lining in Italy and concrete bricks in South Africa (Sims & Huntley, 2004).

Several reports of thaumasite sulfate attack (TSA) are cited in the literature (Hobbs, 2003). In South Africa, TSA by oxidation of pyrites was reported as the consequence of the degradation of concretes blocks containing carbonaceous shale with higher contents of pyrrhotite, pyrite and chalcopyrite. A content of SO_4 of 4.3% by mass was estimated. In other case, in Italy, concrete of 40-years old was diagnosed with thaumasite attack, in which the rock surrounding the structures could be used as aggregate; in damage areas, thaumasite and calcite were present but ettringite was absent. In this case, 1.82% MgO was found in the sound concrete, indicating the presence of dolomite. Thaumasite attack was also observed in 63-years old concrete from Marbjerg Waterworks, in Denmark (Eriksen, 2003). In this case, sulfuric acid attack at the surface was identified as the first part of the attack, by the oxidation of hydrogen sulfide and converted to sulfuric acid by aerobic bacterial action.

From the analysis of field cases in the UK, the Thaumasite Expert Group has identified local potential sources of sulfate leading to thaumasite form of sulfate attack (Clark & BRE, 2002): Kimmeridge Clay, Rheatic Mudstone, and railway ash. In a different way, although thaumasite attack has been found in mortar exposed to sea water, there is no evidence for concern about thaumasite attack in structural concrete exposed to sea water (Clark & BRE, 2002). The main reason seems to be that fulfilling of design parameters for durability against reinforcement corrosion implies a low porosity concrete that is able to resist sulfate attack, and if these precautions are implemented, reinforcement corrosion leads to failure much faster than sulfate attack.

In the case of exposure to $MgSO_4$, the sulfate attack is accompanied by the action of the cation. Conversion of calcium silicate into calcium magnesium has been reported for samples taken in the field (Brown & Doerr, 2000), in agreement with previous laboratory results (Gollop & Taylor, 1992). This is however different from the case of chemical attack when exposed to seawater, where also simultaneous action of magnesium and sulfate takes place, but at a lower concentration of these ions. When exposed to soil containing high amounts of magnesium and sulfate, the combined action allowed magnesium to penetrate further into concrete and form Mg-rich phases after being transported through the augmented pore structure. The deterioration process is therefore much faster. However, subsequent studies carried out by Brown and Hooton (2002) concluded that concretes immersed in Na_2SO_4 solution were more degraded that those in $MgSO_4$ solution. This was attributed to the fact that reaction products containing Mg reduce surface permeability. Also, a reduction in the permeability of concrete was inferred by Brueckner et al. (2012) as consequence of the precipitation of calcite in concrete pores, due to an excess of carbonate in the solution of immersion. The author found that concrete specimens immersed in sulfate solution were more degraded than those immersed in sulfate-carbonate solution. When comparing the progress of deterioration in both aggressive solutions, a reduction of the deterioration rate between 13% and 40% was observed for specimens in SO_4-CO_3 solution.

3.4.3 pH

A correlation between an increasing sulfate content and pH decrease towards the exposed surface observed in tunnels (Fig. 3.1) is interpreted in Lei et al. (2013) as a result of a reduction of the basicity of concrete due to gypsum precipitation in accordance with Eq. 3.1, followed by decomposition of C-S-H according to Eq. 3.2. Therefore, local reduction in pH is caused by the sulfate attack, but if the pH of the surrounding environment is already low, decomposition of C-S-H will take place much earlier due to the previous consumption of portlandite. It has been reported evidence in the field showing that such a drop in the pH is able to cause dissolution

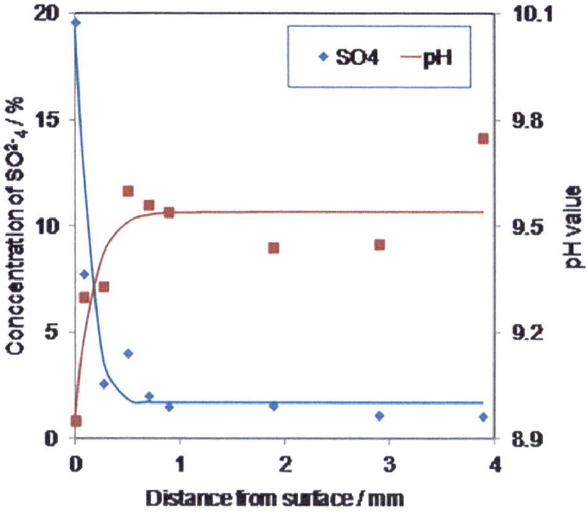

Fig. 3.1 Sulfate and pH profiles in concrete from a railway tunnel exposed to sulfate-laden environment. (Adapted from Lei et al., 2013)

of limestone and dolomitic aggregates that lead to an increase in porosity in the very altered most external millimeters (Leemann & Loser, 2011). This process seems to be linked with the thaumasite form of sulfate attack, and it does not occur where only ettringite is formed. According to Hagelia et al. (2001), the process involving TSA-carbonation is linked with a decrease in pH from 13 to 7, with the requirement of presence of sulfate, carbonate, moisture and low temperature. Moreover, water infiltrations favors the TSA-carbonation process (Hagelia et al., 2003).

$$SO_4^{2-} + Ca(OH)_2 \rightarrow CaSO_4 \cdot 2H_2O \downarrow \quad (3.1)$$

$$C-S-H + SO_4^{2-} + H_2O \rightarrow CaSO_4 \cdot 2H_2O \downarrow + SiO_2 \; or \; Si \; gel \quad (3.2)$$

Microstructural observations of concrete cores suffering sulfate attack have shown that thaumasite starts forming from the center of pores and not from the pore walls (from which sulfate ions are introduced) (Sahu et al., 2003). The authors explain this mechanism of thaumasite formation by the hypothesis of thaumasite forming at a very narrow pH range of about 10–11. Once formed under these conditions, thaumasite seems to remain stable at higher pH levels (Gaze & Crammond, 2000). Three zones are therefore defined according to the influence of pH: an external slightly acidic of slightly basic at the very surface, followed in depth by a region in which thaumasite and gypsum coexist at a pH 1011, and a third region in which thaumasite and ettringite coexist at a pH of 11–12 (Sahu et al., 2003). This would also define the progress of the thaumasite form of sulfate attack with a different dynamic than that of ettringite or gypsum form of sulfate attack, for which crack patterns are formed in the altered zone due to the generation of internal pressures. An example of this can be found in the analysis of a ventilation shaft (Leemann & Loser, 2011), where alteration to thaumasite was found in an up to 25 mm layer from the rock as source of sulfate which showed no cracking related to this alteration. Instead, it seems that the expansion caused by these products caused an increase in stress on the surface and subsequent spalling. In this sense, it seems doubtful whether sulfate attack can be regarded as a surface deterioration process for this case, as it seems that the action of sulfate on a surface of the lining has practical consequences on the opposite face.

3.4.4 Sulfate Concentration

Drimalas et al. (2011) carried out a survey on affected structures in northern Texas. Districts of El Paso, Paris, and Childress were identified as the most likely for sulfate attack, and structures in these areas were inspected. Most of the analysed structures show very little amounts of reaction products (ettringite and gypsum), suggesting that deterioration due to physical attack of salts or some other unidentified process was the main process conducing to damage in the structure. These structures correspond to the Districts of Paris and Childress, where sulfate content in soils was detected in contents lower than 0.08% according to ASTM C 1580, which constitutes a Class 0 exposure according to ACI 201.2R. Four non-significant structures could be connected with sulfate attack in the El Paso District. These corresponded to three concrete riprap slabs and right of way markers. All of them consisted in high water/cement ratio concrete mixes. The sulfate content in the soil was 0.07% as per ASTM C 1580 and 0.18% SO_4 as per TEX-145-E, 0.1% SO_4 as per ASTM C 1580 and 0.15% SO_4 as per TEX-620-J).

One of the riprap showed significant distress, with loosening of aggregate and separation of a top layer in some parts, even when it was exposed to a soil with only 0.01% of SO_4 according to ASTM C 1580 (0.06% according to TEX-145-E). In this case, samples of concrete showed contents of 0.06% and 0.14% at the base and the top of the embankment. This result is explained by a process of concentration of sulfate on the surface of the concrete further up the embankment, due to running water removing the sulfates off of the concrete on the bottom section. Therefore, the high porosity of concrete seems to allow a concentration of sulfate in one part of this particular structure, even when the sulfate concentration in the soil is low. This wick action should however have conducted to damage due to crystallization of salts and not chemical attack. The other structures showed sulfate contents in soil near 0.1% according to ASTM C 1580, which corresponds to the limit between Class 0 and Class 1. It should be emphasized that only concrete of the riprap (with a high water/cement ratio) was affected in these sites, where the main structural elements corresponding to concrete bridges were showing no damage.

In many cases, the concentration of the surrounding environment is of secondary importance, as convection processes such as wick action may concentrate sulfate in particular areas leading to localized damage. An example of this is shown by the assessment of cementitious mortar in a brickwork structure (Crammond, 1985), where damaged mortar contained more than 50% thaumasite by weight of cement whereas mortar in non-damaged sectors of the same structure contained up to 15% of thaumasite+ettringite by weight of cement, which was unable to cause deterioration. However an overestimation of these contents is presumed due to the overlapping of peaks of thaumasite and ettringite influencing the quantitative DRX, a threshold for thaumasite content between 15% and 35% is estimated, and it is therefore worthy to make a comparative analysis. In this case, the source of sulfate were bricks and a similar potential supply of sulfate was available all over the structure, but only in those sectors where convective action lead to a concentration of sulfate chemical attack was found.

Drimalas et al. (Drimalas et al., 2011) point out the difference in the analysis of sulfate content according to different standards. Whereas ASTM C 1580 applies a water/sample ratio of 8:1 to extract sulfates, TEX-145-E and TEX-620-J apply extraction ratios of 20:1 and 10:1, respectively, which lead to higher values for the sulfate content in the soil. This difference seems to be noteworthy as results as per TEX-145-E and TEX-620-J seem to be better connected with the severity of damage and the correspondence with exposure classes as per ACI 201.2R. Actually, it is shown in Drimalas et al. (2011) that the sulfate content in the soil is very dependent on the extraction ratio, and ratios of 20:1 and 10:1 corresponded better than 8:1 to the deterioration observed in the inspected structures.

In Lei et al. (2013), a typical concentration of Na_2SO_4 + $CaSO_4$ in the underground water of 3378 mg/l showed to lead to significant damage due to crystallization of salts in railways tunnels. In another case, a much higher concentration of 6630 of g of SO_4/l has been connected with severe damage due to chemical attack in a tunnel of a dam in China (Ma et al., 2006). Some of the damage could be related to crystallization of salts due to these areas are located in the zone where wick action takes place. However, a deterioration process reaching up to 100 mm depth makes impossible to assign all the damage due to physical attack. At such a depth, the precipitation of salts due to the evaporation of water is very improbable as not only the required permeability to achieve these conditions is extremely high even for low strength concrete, but also as the air in the tunnel is not dry enough (RH > 85%), although wind in the tunnel might be favorable for this process. The chemical attack in this case was confirmed during the inspection, when some concrete that was permanently in contact with the aggressive groundwater could be easily crumbled by hand as cement paste transformed into a white, pulpy mass with no strength (Ma et al., 2006). This deterioration occurred in a relatively short period between the application of new concrete during a reparation and the inspection 8 years later.

In the case of a bridge foundation in the UK, disturbance of the ground has been connected with damage due to sulfate attack by a clay containing up to 5% of pyrites (Hobbs & Taylor, 2000). The weathering seems to have changed the aggressiveness of the soil, as the mobility of the groundwater in undisturbed clay is low. Then, leaching and sulfate penetration are limited. Therefore, the sulfate concentration of the groundwater leading to significant attack seems to be in relation with the permeable properties of the ground, a parameter difficult to consider in laboratory experiments. In addition, the aeration of pyrites causes their oxidation, producing ferric oxides and sulfuric acid (which later neutralizes with calcite and forms calcium sulfate), increasing aggressiveness (Clark & BRE, 2000a, b). In the case there is not enough carbonate in the ground for this neutralization, the oxidation of pyrite reduces the pH and causes a combined attack to concrete (Hobbs & Taylor, 2000).

In vertical structures the aggressiveness of environment may vary due to the variation in the concentration of sulfate in groundwater. An extensive vertical structures studied in Leemann and Loser (2011) showed that the concentration of sulfate increased with depth, from 0.8 to 4.2 g SO_4/l between 12 and 48 m depth. The highest concentration is very significant, although much lower than the usual concentration used in accelerated laboratory experiments (30–50 g/l). This variation in depth also highlights the importance of local conditions regarding the exposure of the structure.

3.4.5 *Other Aggressive Conditions: Carbonation, Chlorides, Freezing and Thawing, Drying Shrinkage*

Field suggestions of the promotion of thaumasite formation with the presence of chloride are presented in Ma et al. (2006). Here, a very high concentration of chloride in the groundwater did not result in formation of Friedel salt, as this is unstable in presence of sulfate and allow a promotion by chlorides of thaumasite formation instead.

In locations where frost action is possible, many times is difficult to differentiate the extent of damage caused by frost action and the amount of damage that can be attributed to sulfate attack (Crammond, 1985). These two deterioration processes work together for enhanced damage, allowing penetration of water and a high saturation degree that potentiates both freeze and thaw and sulfate attack. Therefore, the effect of freeze and thawing on sulfate attack cannot be quantified, and likewise in the opposite way.

Chloride penetration into concrete is favored by both carbonation and sulfate attack. Chlorides are generally partially bound in the cement matrix, but a drop in pH caused by carbonation makes Friedel's salt to be unstable and chloride is released to the pore solution (Brown & Doerr, 2000). For the particular case of the presence of sulfate, the formation of ettringite consumes aluminates that are no longer available for chloride binding. When thaumasite is formed due to further provision of sulfate and CO_3^{2-}, also C-S-H is consumed with less amount of retention capability for the penetration chloride. Then, sulfate attack reduces the resistance of concrete to chloride penetration in the field (Hobbs & Taylor, 2000).

In reinforced concrete, chloride and carbonation produce depassivation of rebars and subsequent corrosion when oxygen and moisture are also available. At this stage there is no significant influence of these processes on sulfate attack, but as corrosion progresses, the formation of expansive oxides at the steel/concrete interface originates tensions and eventually concrete cracks. These cracks allow a faster ingress of chloride, CO_2 and also sulfate, which is generally the slowest of the three. This is a very common situation in reinforced concrete structures exposed to wetting and drying with seawater, where corrosion is much faster than sulfate attack is cover concrete is not protective enough of the reinforcement (Clark & BRE, 2002).

Sulfate attack problems were detected in three rendered parapet wall on buildings in the southeast of England by BRE (Crammond & Dunster, 1997 in Crammond, 2002a, b). The walls were made with sulfate-bearing clay bricks, covered with rendered mortar. Drying shrinkage of the render coats caused the formation of cracks, permitting the ingress of water to the brickwork substrate. After 5 years of application of the render, the walls were sternly damaged by TSA, as well as the adjoining bedding mortar between bricks. Similar deterioration problems were noted by Crammond (1985) in South Wales Buildings.

3.4.6 Temperature

The formation of thaumasite has been repeatedly reported to be promoted at low temperatures (<15 °C). Thaumasite is most favored at temperature of 5 °C, which can be explained because at lower temperatures thaumasite is more insoluble, $Si(OH)_6$ groups are more stable as consequence of an increase in coordination number (Bensted, 1988), and CO_2 solubility and portlandite are increased (Hu et al., 2004 in Shi et al., 2012). However, thaumasite formation in relatively warm locations has been reported (Sahu et al., 2002; Sarkar & Little, 1998; Brown & Doerr, 2000). Thaumasite could not be produced in the laboratory at reaction temperatures above 25 °C (van Aardt & Visser, 1975), then it is derived that this thaumasite is formed during low temperature seasons and prevails partially later. In this sense, Shi et al. (2012) reports that in several laboratory studies the presence of thaumasite is observed at temperatures between 20 °C and 25 °C, and even at 40 °C, indicating in one of them (Pauri & Collepardi, 1989 in Shi et al., 2012) that thaumasite decomposed at temperatures of 60 °C. However an increasing proportion of thaumasite form of sulfate attack is reported since the 1980s than in earlier literature, the identification is complicated, as thaumasite shows very similar XRD patterns as ettringite and probably in many cases it was misinterpreted in the past (Crammond, 1985). Also thaumasite and ettringite are difficult to be differentiated in polished section under the microscope, where only EDS analysis allows to determine the absence of aluminum and confirm the thaumasite phase (Brown & Doerr, 2000).

The ambient temperature is a parameter to be addressed, but microclimatic conditions also need to be considered. For example, in a tunnel in a dam there is a continuous evaporation of water on the surface of the wall, which continuously removes heat from concrete and reduces the temperature with the aid of the isolation that the rest of the structure provides, this might cause temperatures of 4–10 °C in the tunnel (Ma et al., 2006) when the temperature outside is in the wider range of

−10 and 29 °C. A similar situation might explain the detection of thaumasite in a tunnel located in an environment with temperatures of 20–25 °C (Ma et al., 2012). In this case, thaumasite was identified even when external conditions are not completely corresponding to permanent low temperature.

3.4.7 Saturation Degree

The wick action is the main mechanism through which crystallization of salts cause damage in concrete. This is due to the pressure caused by growing crystals in pores, and it is also related to a variable level for the capillary rise when concrete is exposed to a variable water level. Thus, the variable saturation degree causes and increase with time of salt content, reaching extremely high values that led to crystallization. An example of this mechanism is the observed damage in more than 10 existing tunnels in the Chengdu–Kunming railway, China (Lei et al., 2013). Therefore, it is the variation in the saturation degree more than the saturation degree itself that can be connected with the level of deterioration. Moreover, it should be considered the variation in the moisture content, as significant wick action takes place in concrete with an extended network of capillary pores. This is, concrete made with a low water/cement ratio and a consequent low porosity will reach a high saturation degree for a small amount of liquid, but the variation in time will be much slower due to limited transport properties of the material, and the amount of salts introduced with each cycle of wetting is also smaller.

The service conditions of tunnels are therefore especially suitable for the promotion of crystallization of salts that contribute to damage together with chemical attack. Physical attack has been reported in all cases (Ma et al., 2006, 2012; Lei et al., 2013), and especially railway and road tunnels add on top the effect of negative pressure of the circulation of traffic to produce further evaporation on the surface of concrete. The combined action in tunnels is schemed in Fig. 3.2. Water containing sulfate is permeated through concrete and especially joints, and the evaporation on the surface leads to the concentration of salts. Also drainages contribute by supplying water that ascends by capillary absorption through the walls and evaporates at a higher level. The wind and negative pressure caused by traffic helps for the increasing of the evaporation rate. A permanent moisture gradient is generated, which allows a continuous supply of salts towards the inner surface of the concrete structure. This surface is the sector where most of damage is founded in the field, at low levels on the walls, side drainages and proximity of constructive joints and structural cracks.

Severe damage in the tunnel of a dam seems to be related to a high evaporation rate at the surface (Ma et al., 2006). This process promotes the crystallization of salts. In this particular case, due to the nature of the structure and the convection effects causing wind in the tunnel intensifies the moisture gradient in the exposed

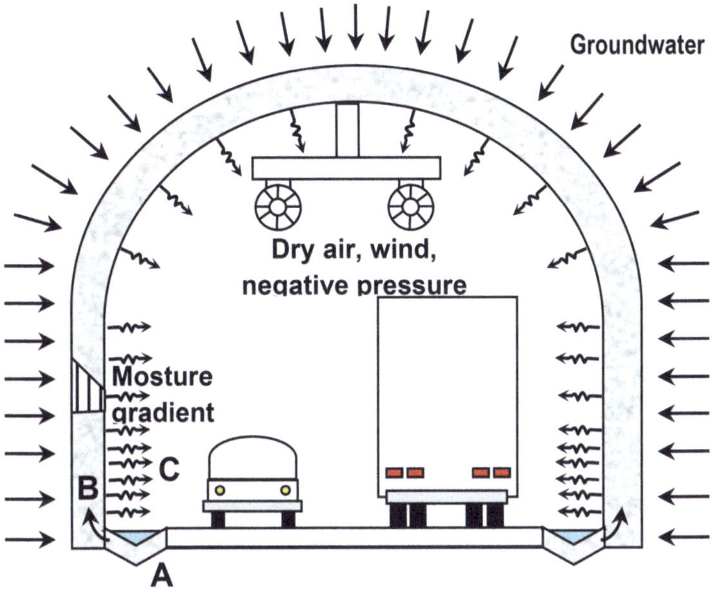

Fig. 3.2 Transport mechanisms contributing to crystallization of salts and sulfate attack in tunnels. (**a**) Drainages with permanent salty water supply; (**b**) capillary absorption; (**c**) evaporation

concrete. With a greater moisture gradient, a thicker layer of concrete is expected to be affected by the physical deterioration mechanism.

There is some suggestion that bituminous waterproofment might prevent the onset of the thaumasite form of sulfate attack form (Clark & BRE, 2002). This favorable practice requires further evidence, but it is possible that it is linked to the prevention of moisture transport in concrete, maintaining a constant saturation degree in the material. On the other hand, if an impervious render or covering is badly applied, water might become trapped and eventually lead to the saturation of the substrate generating ideal conditions for sulfate attack (Crammond, 1985).

A particular aspect of the influence of the saturation degree might be identified at early age, when premature drying in the material causes cracking due to drying shrinkage, and water is given access through these cracks to facilitate sulfate attack in the mass of the material (Crammond, 1985). In this internal interfaces water remains longer and the saturated material is affected by sulfate attack with a consequent progressing of this cracking pattern.

The saturation degree can also important in some cases for the investigation of the damaged material. Thaumasite is generally stable when exposed to dry air, but some fine grains of thaumasite can be unstable when the sample is dried out (Crammond, 1985), and less amount than the one causing damage might be detected experimentally.

3.4.8 Cement Type

In several countries, Portland cements can contain up to 5% of limestone filler as minority component. The presence of limestone in cement allows the formation of thaumasite, as carbonate is a necessary component for this process. Although the Thaumasite Expert Group indicated that contents of limestone filler as low as 5% will not affect concrete performance, some studies have concluded that thaumasite can form in cement paste made with only 5% limestone filler (Sims & Huntley, 2004).

A particular case of damage of concrete in a tunnel of a dam show that even when the use of a low dose of cement with low C_3A was used limited the amount of ettringite that could be formed, specific temperatures and humidity (<10 °C, RH >85%) leads to C-S-H to react with sulfate in very high concentration in the water and carbonate to form thaumasite (Ma et al., 2006), this seems to be in contradiction with early studies that indicate improbable formation of thaumasite in absence of alumina in the form of ettringite in the system to initiate the growth of thaumasite (Crammond, 1985; Sahu et al., 2003). On the contrary, low C_3A cements seem in practice unlikely to fully provide protection against this particular form of sulfate attack. In agreement, deterioration due to thaumasite formation has been identified in a number of highway structures in the Gloucestershire, Somerset and Wiltshire (UK), with most of them containing OPC, but also some of them containing sulfate resistant cements (Clark & BRE, 2002), as well as a ventilation shaft in Switzerland made with low C_3A (less than 2%) cement (Leemann & Loser, 2011). Different study cases on sprayed concretes in Norway showed that the content of C_3A in the cement seem to not be related with the degradation mechanism of TSA-Carbonation. For these cases, C_3A content varied between 1% and 8% (Hagelia et al., 2003).

From the observation of case studies, the Thaumasite Expert Group recommends that Portland Limestone Cements (up to 35% of limestone powder content) should not be permitted for constructions in contact with groundwater containing more than 0.4 g/l of sulfate. The use of Ordinary Portland Cements, for which up to 5% of limestone is allowed in Europe, has shown a slightly detrimental effect for sulfate resistance in cold weather in laboratory studies (Hartshorn et al., 1999), but this seems insufficient as no evidence of this small affectation could be collected for structures in the field.

From comparative studies for the performance of different binders against the thaumasite form of sulfate attack (Crammond, 2003) have shown dissimilar convenience of different mineralogical composition and supplementary cementitious materials. Portland cement with C_3A contents of 7–10% is not resistant to thaumasite sulfate attack neither in the field or laboratory tests. Sulfate-resistant Portland cement shows improvement over high-content C_3A cements in tests conducted in the laboratory for 2 years. Brown and Hooton (2002) have found that the use of moderate-resistant Portland cement (Type II, ASTM) considerably increases the depth of sulfate attack of concretes exposed to two different attacking species ($MgSO_4$ vs. Na_2SO_4), in comparison with concretes made with sulfate-resistant Portland cement. The different performance could be also associated with a higher porosity of concretes with moderate-resistant cement (w/c = 0.50) respect to those with sulfate-resistant cement (w/c = 0.45). However in the field the performance after 4 and 6 years of exposure resulted in similar deterioration for concretes with both types of cement. Inconsistencies between results from the laboratory and the field have also been found for the effect of fly ash, which is reported to be beneficial in the field but not in the laboratory. Beneficial effects of silica fume and metakaolin have been verified in the field, but not tested in the laboratory. By its side, the inclusion of high replacement ratios by slag consistently showed excellent results in the field and laboratory. In this sense, Hobbs (2003) concluded that concretes made with 70% slag cement presents an enhanced performance in the presence of magnesium ions and carbonate aggregates, even at 5 °C, whereas SRPC concretes show a reduced performance under these conditions. In the same way, concretes prepared with different types and content of SCMs no present signs of damage after 3 years exposure to sulfate solution. The better performances were noted for concretes with 40% and 50% slag, 25% slag+10% metakaolin, 25% slag+6% silica fume, and OPC + 8% silica fume. Concretes with 15% metakaolin+40% slag showed low damage (Hooton et al., 2013).

3.4.9 Aggregates

In an extensive survey of highway structures in the UK, 16 cases of thaumasite form of sulfate attack were identified from a total of 733 structures (Clark & BRE, 2002). All the affected elements have contained high carbonate limestone aggregate as constituent, which constitutes a significant association between thaumasite attack and limestone aggregate. Two exceptional cases are also reported in Clark & BRE, (2002), in which affected concrete contained low carbonate siliceous gravel, supporting the possibility that concretes containing little or no carbonate could be affected by thaumasite attack when an external source of carbonate ions is available (groundwater). Provision of carbonates by groundwater is also inferred in Sahu et al. (2002), where thaumasite formation is also identified in absence of limestone. Early studies have confirmed that only small amounts of thaumasite are able to form when siliceous aggregates are used and no external supply of CO_3^{2-} is provided, even after 4 years in autoclave at 5 °C, but dolomitic aggregate mortars exposed at the same conditions disintegrated into soft pulpy mass containing substantial amount of thaumasite (van Aardt & Visser, 1975).

Regarding the precipitation of reaction products, the interfacial transition zone seems to be a preferential region. The altered zone of concrete lining next to sulfate laden rock showed that thaumasite formed predominately along the interface between aggregates and cement paste, whereas its presence in air voids was rarer (Leemann & Loser, 2011). It could be derived that the texture of aggregate and the physical-mechanical compatibility between aggregate and cement paste might be a significant aspect regarding dynamics terms of the process.

However, the presence of carbonate aggregate does not necessary conduct to thaumasite formation or thaumasite sulfate attack. Limestone aggregate concretes used in several concrete structures in France have not shown evidence of TSA or thaumasite formation in many decades. The explication of that is a high content of blast-furnace slag in the cements used for make these concretes (Sims & Huntley, 2004).

3.4.10 Mechanical Loads

An investigation of a vertical ventilation shaft exposed to sulfate-containing groundwater for 45 years in Switzerland (Leemann & Loser, 2011) showed no significant correlation between compressive strength of cores and the degree of damage they presented at the moment of the analysis. The authors explain this lack of correlation by the nature of external chemical attack, which is a surface deterioration process in contrast with compressive strength which is a volumetric property. It can be therefore supposed that correlation between mechanical load and the level of damage is not very direct. A possible linkage is the contribution that mechanical load of compression could provide to facilitate spalling, this is not possible to be verified in this field study, as the stress at which concrete was submitted in service in the different locations is unknown.

3.5 Soils as an External Source of Sulfates

One of the most common sources of sulfates are soils with a high sulfate content. Soil may contain excessive amounts of gypsum or other sulfates from the oxidation of sulfide minerals (pyrite, marcasite, pyrrhotite). These can penetrate the cement matrix. Soils containing dihydrated calcium sulfates are usually located in dry areas.

Acid sulfate soils suffer extreme acidity as a result of oxidation of pyrite. Often they are also unripe; sometimes also saline. Some occur naturally but most have developed as a result of drainage of previously waterlogged coastal alluvium and peat. Acid sulfate soils pose a range of problems for communities dependent on the reclaimed land—including low crop yields, a restricted range of alternative uses, soil engineering hazards, water pollution, and other environmental risks. Civil engineers must cope with a unique combination of corrosion by acidity, salt, and reducing conditions, and with the difficulties of design and construction of earthworks, roads, and drainage systems in unripe materials (Dent, 1986).

Acid sulfate soils develop as a result of the drainage of parent materials that are rich in pyrite, FeS_2. Pyrite accumulates in waterlogged soils that are both rich in organic matter and flushed by dissolved sulfate, usually from sea water. When drainage brings oxygen into these previously waterlogged soils, the pyrite is oxidised to sulfuric acid. Acid sulfate soils develop where the production of acid exceeds the neutralising capacity of the parent material, so that the pH falls to less than 4. Acid sulfate soils are an almost unique case where the soil problems are so severe that they can dominate most other aspects of land development: from engineering works (including the kind of concrete or steel required, design of roads, embankments, and drainage systems), to agricultural systems (including the choice of crops, disease, lime and fertilizer requirements), to economic and social planning at regional and local level, to the environmental impact of reclamation. On a world scale, acid sulfate soils are not extensive. But they are important in many regions of critical population pressure, notably in South East Asia and West Africa, where alternative land for subsistence food production and cash crops is not available. The tidal swamps offer apparently attractive and easily reclaimed land, but the distribution and severity of the acid sulfate hazard cannot be assessed without detailed soil survey. The soil problem is deceptively simple-excess acid production leading to toxicity; and the slow rate or, alternatively, high cost of amelioration. However, acid sulfate soils are not uniformly and equally bad; often they occur in intricate patterns in association with non-acid soils. Always there are local variations in the nature and severity of problems, and in their response to alternative management practices. Better management requires (Dent, 1986):

- Soil survey to identify and so avoid, or at least anticipate, acid sulfate problems;
- A code of practice for field experimentation and site characterisation, so that experience and experimental data can be transferred to similar areas;
- Quantitative data as a basis for decisions about land development. These may be provided by quantitative models to predict the rate of acid generation and leaching following drainage, the rate of pH rise following flooding, the extent of iron and other toxicity problems, and the rate of amelioration of acid sulfate soils under alternative management;
- Practical guidelines for the reclamation and management of acid sulfate soils, in particular for the identification of the problem in the field, and simple cheap measures to minimise the oxidation of pyrite and combat its consequences;
- Long-term monitoring of alternative management practices and local trials in cooperation with the farmers working the land. These will provide a basis for introducing new practices.

Dent (1986) also comments on its publication that cement is attacked by dissolved sulfate; steel is subject to slow dissolution by strong reductants; both steel and cement are rapidly attacked by acid. Each of these conditions may occur in acid sulfate areas. Several approaches are possible to extend the life of steel and concrete structures in these circumstances:

- Avoidance: The distribution of aggressive soils must be mapped. It may then be possible to site structures in safe locations where acid sulfate conditions will not occur. Avoidance will be the cheapest solution whenever it is feasible.
- Tolerance: Materials that resist acid sulfate conditions, such as stainless steel, sulfate-resistant cement, acid-resistant masonry or hardwood, can be used. Generally this will be a very high-cost solution. Stainless steel would be consid-

ered only for key parts of costly structures. Specifications for tolerant materials must take account of the degree of acidity and salinity expected. In acid sulfate soil, pH values may locally be as low as 2 for long periods. Dissolved sulfate contents in acid sulfate soils range from 10^3 to 2×10^4 g m^{-3}, compared with 3 to 4 g m^{-3} in average river waters and 3×10^3 g m^{-3} in seawater. Minimising the area of contact with aggressive solutions is often practicable. Porous concrete must be avoided. The sand and gravel aggregate must not contain salts. Smooth, pre-cast piles are preferable to piles poured in place.
- Sacrifice: A thickness of sacrificial concrete may be included in the design of structures; steel reinforcement must be covered with at least 50 mm of concrete. Further protection may be achieved by mixing lime with the soil layer surrounding the metal or concrete. This requires an excess of lime over the potential acidity of the soil (for example 5% by mass, in a layer about 1 m thick around the structure). The lime barrier may not protect indefinitely where there is lateral movement of acid groundwater. Metals can be cathodically protected from acid attack by sacrificial electrodes.
- Insulation: Susceptible materials can be coated with plastic or bituminous composition. Defects in the insulation may cause local dissolution or failure.
- Inhibition: Acids and sulfates can be reduced to solid sulfides by maintaining water saturation in the surrounding soil. If reduction will not start spontaneously, it can be promoted by mixing in some decomposable organic matter. In some cases, iron oxide may be needed as well. The hazard of dissolution remains until the pH has become near neutral; sulfate attack of cement will persist until the soluble sulfates are reduced. If waterlogged conditions are not maintained, the hazards will recur.

All conventional concretes are vulnerable in varying degrees to attack by inorganic salts and inorganic/organic acids present in groundwater. In soil and natural groundwater, sulfates of sodium, potassium, magnesium, and calcium are common. Sulfuric acid, when present in natural groundwater, forms from the oxidation of iron sulfide (pyrites) minerals in soil (Hobbs & Taylor, 2000).

Peat soils, clay soils, and alum shale can contain iron sulfide (pyrite) which, upon oxidation, produces sulfuric acid. Further reaction can produce sulfate salts, which produce sulfate attack (ACI, 2001).

The external sulfate attack induced damage, which is the traditional sulfate attack, is determined by the chemical interaction of a sulfate-rich soil or water with the cement paste. Soils containing sodium, potassium, magnesium, and calcium sulfate are the main sources of sulfate ions in groundwater. For external sulfate attack to occur, the following three conditions must be fulfilled (Collepardi, 2003):

- high permeability of concrete;
- sulfate-rich environment;
- Presence of water.

Roads and Traffic Authority (RTA, 2008) in their report comment that Acid Sulfate Soils, which include both Actual Acid Sulfate Soils and Potential Acid Sulfate Soils, have been identified in many parts of Australia, with consequent effects on the environment and structures. For many design and construction engineers and project managers, the knowledge, implications and actions required to deal with Acid Sulfate Soils are not well understood.

Actual Acid Sulfate Soils contain highly acidic soil horizons or layers resulting from the oxidation of soil materials that are rich in iron sulfides. Oxidation produces hydrogen ions in excess of the sediments capacity to neutralise this acidity, resulting in soils with pH of 4 or less when measured in the dry season.

Potential Acid Sulfate Soils contain iron sulfides or sulfidic material that has not been exposed to air and oxidised. The measured pH of these soils in their undisturbed state is usually 4 or more (and may be neutral or even slightly alkaline).

Acid Sulfate Soils includes both actual Acid Sulfate Soils and potential Acid Sulfate Soils. Both types may be found in a soil profile.

When water passes through actual Acid Sulfate Soils, sulfuric acid is leached out to react with the minerals in the soil to change its properties. If the soil has insufficient buffering capacity to neutralise the acid, the soil water, groundwater and drainage water will all become acidic, and will contain dissolved aluminium, iron and heavy metals.

Engineering works on Acid Sulfate Soils, such as excavation, dredging and draining accelerate the exposure to air of pyritic soils i.e. those that contain natural sulfide, otherwise known as iron pyrites, (FeS_2). These operations can speed up the production of acidic groundwater to many times the natural rate.

Passing into waterways and groundwater, the sulfuric acid produced as a result of soil disturbance affects plant growth, aquatic life, animal and human health, and degrades engineering structures.

In addition to the risk of deterioration of engineering structures in actual Acid Sulfate Soils, ASS, unconsolidated estuarine sediments containing potential Acid Sulfate Soils may cause uneven subsidence under relatively low loads, causing structural problems.

Depending on the predominant chemical reaction, concrete deterioration processes can be classified into three groups according to their effects on the hardened cement paste: leaching; exchange reactions and removal of readily soluble compounds; swelling, due to the formation of new, stable compounds. Acids in concentrations common to natural waters can dissolve carbonates at the surface of the concrete and leach out lime. The concrete deteriorates because the calcium hydroxide in the cement paste is attacked by the acids to form water soluble salts subsequently leached out. Sulfuric acids may also cause swelling of the paste with subsequent cracking. The rate of attack on the concrete is controlled by the nature of the acid, the pH value and the solubility of the calcium salts formed.

Sulfate attack on concrete is caused by an expansive chemical reaction between tricalcium aluminate (C_3A) components in the cement and the sulfates in solution to produce gypsum and calcium sulfoaluminate (ettringite). The crystals of ettringite occupy a larger volume than the original cement components that leads to expansion, cracking, and disintegration of the concrete. The aggressiveness of soil containing sulfates is classified according to SO_3 content, more recently according to

SO_4 content. SO_3 can be converted to SO_4 as follows: $SO_3 = 0.83\ SO_4$. Ammonium, calcium, magnesium and sodium sulfates are most detrimental to Portland cement. Potassium, copper and aluminium sulfates are less harmful. Barium sulfate and lead sulfate, which are insoluble in water, do not affect concrete (RTA, 2008).

According to the NSW Acid Sulfate Soils Management Advisory Committee (NSW, 1998), acid sulfate soils are the common name given to naturally occurring sediments and soils containing iron sulfides (principally iron sulfide or iron disulfide or their precursors). The exposure of the sulfide in these soils to oxygen by drainage or excavation leads to the generation of sulfuric acid. "Acid sulfate soils" include actual acid sulfate soils or potential acid sulfate soils. Actual and potential acid sulfate soils are often found in the same soil profile, with actual acid sulfate soils generally overlying potential acid sulfate soil horizons. "actual acid sulfate soils" are soils containing highly acidic soil horizons or layers resulting from the aeration of soil materials that are rich in iron sulfides, primarily sulfide. This oxidation produces hydrogen ions in excess of the sediment's capacity to neutralise the acidity resulting in soils of pH of 4 or less when measured in dry season conditions. These soils can usually be identified by the presence of pale yellow mottles and coatings of jarosite. "Potential acid sulfate soils" are soils which contain iron sulfides or sulfidic material which have not been exposed to air and oxidised. The field pH of these soils in their undisturbed state is pH 4 or more and may be neutral or slightly alkaline. However, they pose a considerable environmental risk when disturbed, as they will become severely acid when exposed to air and oxidised.

Acid sulfate soils can have considerable effects on:

- Engineering and landscaping works (including affecting the type of concrete or steel required, the design of roads, buildings, embankment and drainage system, extractive materials specifications, maintenance programs for drains, water and sewage pipelines and other structures);
- Agricultural and tea tree management practices (including the choice of crops, liming practices, fertiliser requirements, drainage practices, laser levelling);
- Aquaculture management practices (including choice of site, pond design and management practices, liming practices);
- The management of contaminated soil (particularly in relation to mobility of metals);
- The conservation of biodiversity and protection of wetlands and shallow freshwater aquifer systems (in relation to the degradation of habitat or water quality, killing or disease of fish and other aquatic organisms).

The following activities undertaken in areas likely to affect or use coastal sediments, warrant an assessment of the risk of exposing acid sulfate soil:

- Excavation or disturbance of acid sulfate soil: For example in construction of roads, foundations, drainage works, laser levelling, land forming works, flood mitigation works, dams and aquaculture ponds, sand or gravel extraction, dredging. When acid sulfate materials are being excavated, attention must be given to the excavation site as well as the location where the excavated material is placed or used.

- Lowering the watertable: For example in new drainage works or deepening of existing drains, use of groundwater, de-watering of dams, wetlands or quarries, dredging works lowering the bed of a river
- Use of acid sulfate soil For example in aquaculture pond walls, dams, flood mitigation works, imported fill material, reclamation or foreshore works
- Physical habitat modification for mosquito control For example: runnelling, drainage and selective ditching to remove water or allow predatory fish access to tidal pools

In reference to analyse soil and water indicator for to detect an acid sulfate soil, the investigation should include a field inspection to consider soil and surface and subsurface water characteristics and if necessary, limited groundwater analysis. NSW Acid Sulfate Soils Management Advisory Committee (NSW, 1998) suggests different analysis methods for the identification of acid sulfate soils.

As many of the indicators for actual and potential actual acid sulfate soils are quite different, the field inspection should investigate for the presence of both soils. Commonly actual acid sulfate soils are found overlaying potential acid sulfate soils and both are usually covered by non-acid sulfate alluvial topsoil.

If soils or associated waterbodies demonstrate one or more of the following indicators in Table 3.1, it can be assumed acid sulfate soils are present and a full assessment should be undertaken. Soil field pH provides a useful quick indication of the likely presence and severity of "actual" acid sulfate soils (NSW, 1998).

Field pH readings should be taken at regular intervals down the soil profile. pH readings of pH ≤ 4, indicates that actual acid sulfate soil are present with the sulfides having been oxidised in the past, resulting in acid soils (and soil pore water). pH readings of pH >4 may indicate the absence of "actual" acid sulfate soils but does not give any indication as to the presence of "potential" acid sulfate soils.

In reference of groundwater analysis as indicators of acid sulfate soils, the analysis of groundwater or drain water for the soluble chloride: soluble sulfate ($Cl^-:SO_4^{2-}$) ratio can indicate that sulfidic material in the vicinity of the site is being, or has been, oxidised.

As well soil suspensions/slurries can be examined under a microscope for sulfide framboids and individual crystals. As a further confirmation, the reaction of the sulfide to peroxide can be observed on the slide. If effervescence is observed, this provides a positive test to confirm the presence of sulfide. However, failure to see crystals or framboids is not evidence that sulfide is absent as sulfidic crystals may have been lost in the sampling or slide preparation. Caution is required with the use of this technique as it requires acquired skills. False positives are common when high levels of organic material or manganese are present.

According to (TSA, 2008), there are several methods and tests available to identify ASS. Two simple methods of identification are:

Table 3.1 Classification soil type in function of observations soil or water field indicators

Soil type	Indicators
Acid sulfate soils	*Landscape characteristics:*
	• dominance of mangroves, reeds, rushes and other marine/estuarine or swamp-tolerant vegetation
	• low lying areas, back swamps or scalded/bare areas in coastal estuaries and floodplains
	• sulfurous smell after rain following a dry spell or when the soils are disturbed
actual acid sulfate soil	*Soil characteristics:*
	• field pH ≤4 in soils
	• presence of shell
	• any jarositic horizons or substantial iron oxide mottling in auger holes, in surface encrustations or in any material dredged or excavated and left exposed. Jarosite is a characteristic pale yellow mineral deposits which can be precipitate as pore fillings and coatings on fissures. In the situation of a fluctuating watertable, jarosite may be found along cracks and root channels in the soil. However, jarosite is not always found in actual acid sulfate soils.
	Water characteristics:
	• water of pH <5.5 in adjacent streams, drains, groundwater or ponding on the surface
	• unusually clear or milky blue-green drain water flowing from or within the area (aluminium released by the acid sulfate soils acts as a flocculating agent.)
	• extensive iron stains on any drain or pond surfaces, or iron-stained water and ochre deposits
	Landscape and other characteristics:
	• scalded or bare low lying areas
	• corrosion of concrete and/or steel structures
potential acid sulfate soils	*Soil characteristics:*
	• waterlogged soils—unripe muds (soft, buttery, blue grey or dark greenish grey) or estuarine silty sands or sands (mid to dark grey) or bottom sediments of estuaries or tidal lakes (dark grey to black)
	• presence of shell
	• soil pH usually neutral but may be acid
	Water characteristics:
	• water pH usually neutral but may be acid

- visual indicators of ASS: Visual indicators of ASS at a site may include yellow efflorescence on the soil surface, sulfurous smell, iron staining and iron flocculants in streams.
- on-site pH measurements: On-site pH of streams and fresh groundwater samples and of 1:5 soil:water samples provide a good indication of acid sulfate conditions at the site.

Where soils contain enough sulfides, the sulfate content of groundwater in construction pits, wells or boreholes may increase over a period of weeks to several times the original value. After backfilling, the sulfate content soon drops to previous levels, since the supply of air has been interrupted. This explains why water samples taken

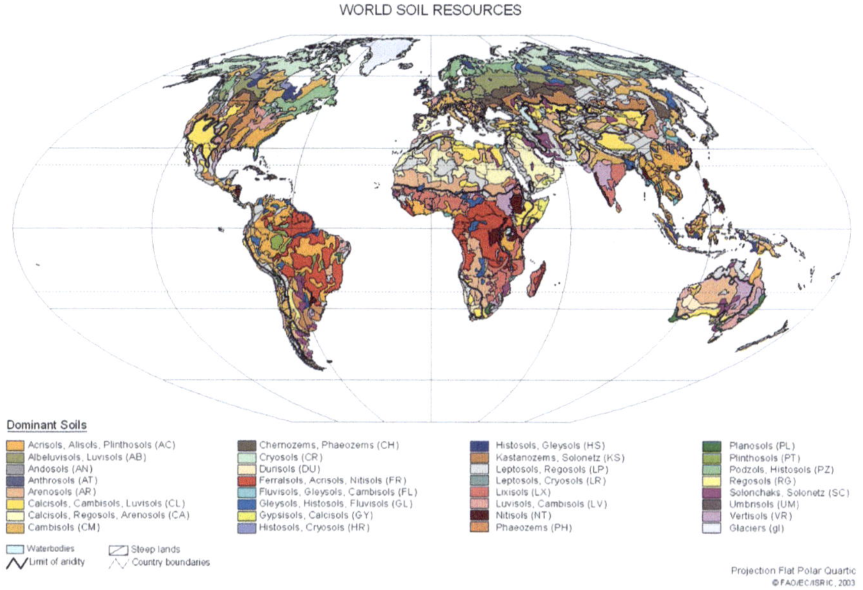

Fig. 3.3 World soil resources (FAO/EC/ISRIC, 2003)

from construction pits are usually higher in sulfates than those obtained from exploratory drilling. Protective measures based on the higher sulfate content of samples taken from construction pits could be excessively conservative.

Moving water is particularly harmful to concrete. In stagnant water, dissolved salts will tend to combine with components of the hardened cement paste. Sodium sulfate in groundwater will react with calcium hydroxide in the hydrated cement to form gypsum. The concrete pores are then partially sealed by the precipitated gypsum to form a natural protective layer on and near the concrete surface. In moving water aggressive acid sulfates may be replenished. In stagnant water the acid sulfates are consumed with time. In cohesive soils such as clay the seepage rate of groundwater is of the order 10^{-5} m/s whilst in granular soils seepage rates a hundred or even a thousand times greater are possible. Therefore, in granular soils higher rates of deterioration should be expected.

In Fig. 3.3 is presented the map of world soil resources (FAO/EC/ISRIC, 2003). The yellow colour areas correspond to the gypsum soils and calcium soils. In an overall assessment, soils containing calcium sulfates and calcium carbonates are located in the west side of North America, North, Central and South Africa, central regions of the Asian continent and south of Australia. In general, these zones coincide with too dry areas, shown in yellow in the map of Fig. 3.4.

In Fig. 3.5 the world map pH in topsoil is presented.

In Figs. 3.6 and 3.7 it is possible to observe on a world-wide level the zones with excessive salts presence, where calcium carbonates and calcium sulfates dehydrate,

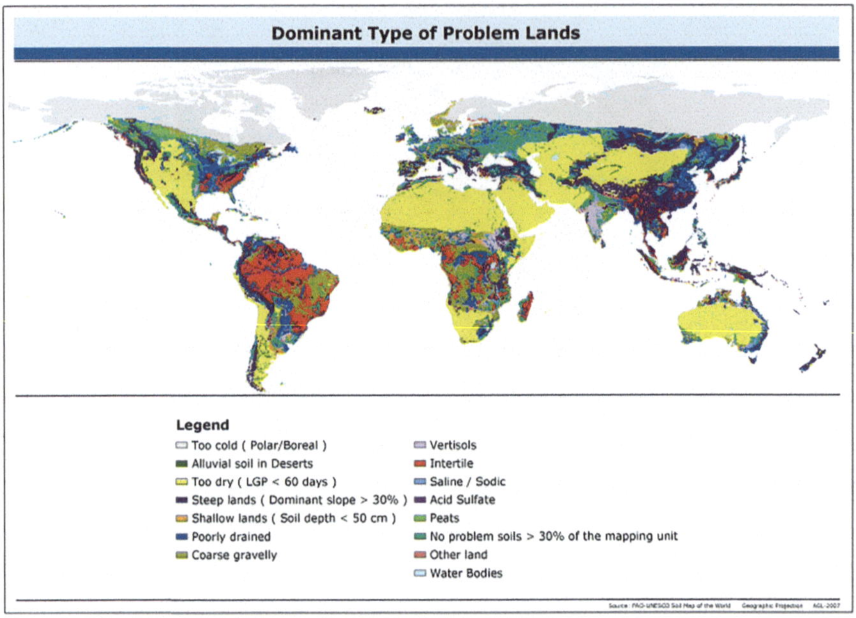

Fig. 3.4 Too dry zones located in world map (FAO-UNESCO, 2007)

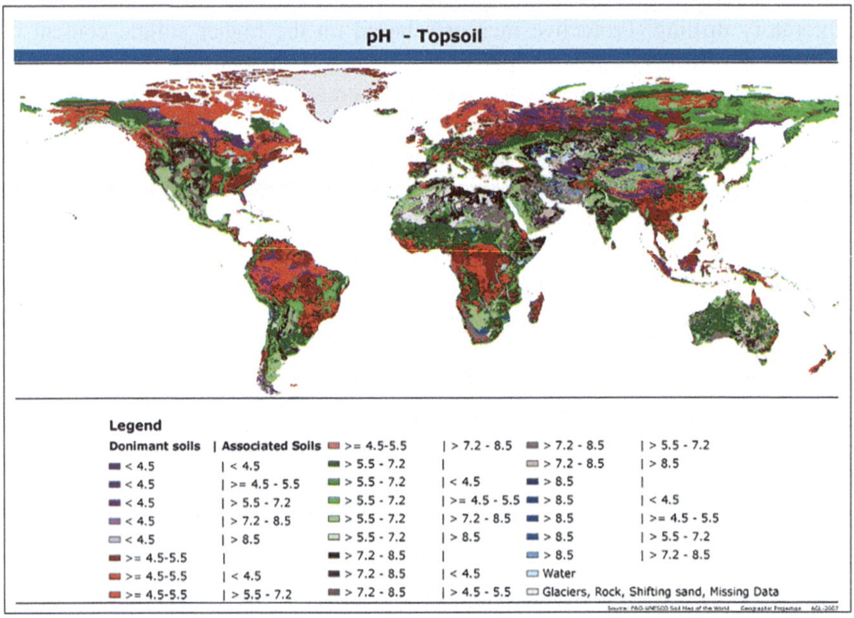

Fig. 3.5 pH-Topsoil located in world map (FAO-UNESCO, 2007)

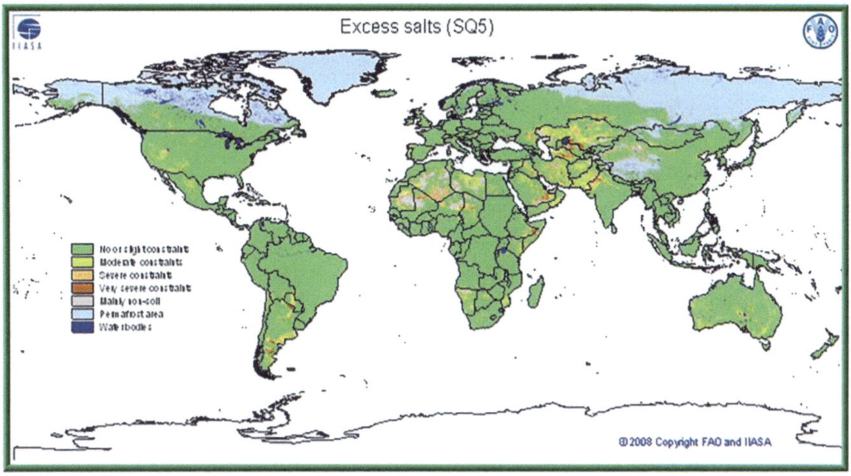

Fig. 3.6 Excess salts in world map (Fischer et al., 2008)

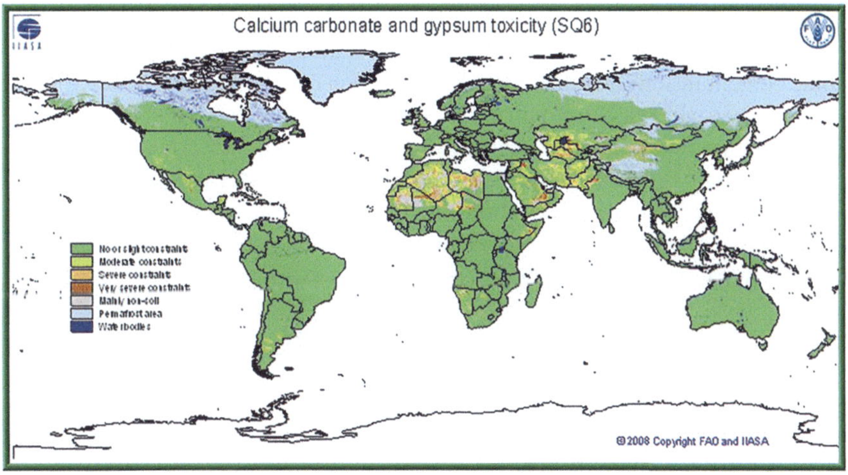

Fig. 3.7 Calcium carbonate and gypsum toxicity in world map (Fischer et al., 2008)

indicated in yellow, orange and brown, being potential zones of external contribution of sulfates.

Sulfates in the hardcore or soil will not be damaging the concrete if they remain dry. Water is necessary to dissolve the sulfates and carry their anions into the concrete. The amount of water needed is not too much. The surrounding soil needs not to be saturated, just be humid. In some cases the moisture content in the soil is just between 12% and 14%. A common source is water from the sub-soil that is carried up through the finer fraction of the hardcore due to capillary action. Once in contact with concrete, water is drawn by capillary action due to evaporation at the surface (Zarzar & Almeida de Oliveira, 2011).

Calcium-absorptive acidic soil can attack concrete, especially porous concrete. Even slightly acidic solutions that are lime deficient can attack concrete by dissolving calcium from the paste, leaving behind a deteriorated paste consisting primarily of silica gel. Sulfate attack is a particular problem in arid areas (PCA, 2002).

Generally, the distribution of sulfates in soil varies both horizontally and vertically. For example, in the U.K., often, the top meter or so of soil is relatively free from sulfates because of leaching by groundwater. On the other hand, in dry regions, where the rate of evaporation is high, there may be a higher concentration near the surface. Drainage of soil around the foundations is important because dry salts do not attack the concrete. The classification of severity by the determination of the sulfate content in the soil is still a matter of debate. The determination of the sulfate content in the soil should be standardized in terms of the extraction ratio, so that the outcome of the test cannot be varied according to the aim or objectives of the tester. The amount of sulfate should be expressed in milligrams of SO_4 per kilogram of extract water (Neville, 2004).

Exposure classes with respect to sulfate concentration in soils and groundwater, according to various international standards

In 2004, Neville made an overview of the methods that exist to determine the sulfate content in soils and that give a classification of exposure to sulfates.

In the next tables a selection of tables from American, Canadian, British, and European sources give a classification of exposure to sulfates. These tables show that it is not only the sulfate content boundaries of the various exposure classes that vary, but also the water–soil extraction ratios (sometimes unspecified), as well as the method of determining the sulfate content: acid or water soluble. Furthermore, the concentration of the sulfate in the soil is sometimes expressed per unit mass of soil; in other cases, per unit mass of extraction water. The rationale of these various approaches is not self-evident (Neville, 2004) (Tables 3.2, 3.3, 3.4, 3.5, 3.6, 3.7, 3.8, 3.9, 3.10, 3.11, 3.12, 3.13, 3.14, 3.15, and 3.16).

In general, Portland cement concrete does not have good resistance to acids. In fact, no hydraulic cement concrete, regardless of its composition, will hold up for long if exposed to a solution with a pH of 3 or lower. However, some weak acids can be tolerated, particularly if the exposure is occasional. Acids react with the calcium hydroxide of the hydrated Portland cement. In most cases, the chemical reaction

Table 3.2 ACI 201.2R (1997) Guide to durable concrete. ACI 332R (1984, reapproved 1997). Guide to Residential Concrete

	Concentration of sulfates as SO_4	
Class of exposure	In groundwater (ppm)	Water soluble in soil (%)
Mild	<150	<0,10
Moderate	150–1500	0,10-0,20
Severe	1500–10,000	0,20–2,00
Very severe	>10,000	>2,00

Table 3.3 ACI 318 (1983 through 2002) Building Code

Class of exposure	Concentration of sulfates as SO_4	
	In water (ppm)	Water soluble in soil (Percent by weight)
Negligible	<150	<0,10
Moderate	150–1500	0,10-0,20
Severe	1500–10,000	0,20–2,00
Very severe	>10,000	>2,00

Table 3.4 U.S Bureau of Reclamation Concrete Manual (1966)

Class of exposure	Concentration of sulfates as SO_4	
	In water (ppm)	Water soluble in soil (ppm)
Negligible	0–150	<1000
Moderate	150–1000	1000–2000
Severe	1000–2000	2000–5000
Very severe	>2000	>5000

Table 3.5 Uniform Building Code (1991)

Class of exposure	Concentration of sulfates as SO_4	
	In groundwater (ppm)	Water soluble in soil (Percent by weight)
Negligible	<150	<0,10
Moderate	150–1500	0,10-0,20
Severe	1500–10,000	0,20–2,00
Very severe	>10,000	>2,00

Table 3.6 Canadian Standard A23.1.94

Class of exposure	Concentration of sulfates as SO_4	
	In groundwater (ppm)	Water soluble in soil (%)
S-3 moderate	150–1500	0,10-0,20
S-2 severe	1500–10,000	0,20–2,0
S-1 very severe	>10,000	>2,00

Table 3.7 European Standard EN206-1:2000

Class of exposure (chemical environment)	Concentration of sulfates as SO_4		Concentration of magnesium in groundwater (ppm)
	In groundwater (ppm)	In soil (ppm total)	
Slightly aggressive	200–600	2000–3000	300–1000
Moderately aggressive	600–3000	3000–12,000	1000–3000
Highly aggressive	3000–6000	12,000–24,000	3000 to saturation

The extraction of SO_4 from soil is by hydrochloric acid. However, water extraction may be used if experience is available in the place of use of the concrete

Clay soils with a permeability below 10^{-5} m/s may be moved into a lower class

Table 3.8 British Building Research Station: notes on concrete in sulfate-bearing clays and groundwaters, The Builder, July 7, 1939, p. 29. This table was incorporated into British BRE Digest 31 (1951)

	Concentration of sulfates as SO_4	
Class of exposure	In water (ppm)	In clay (Percent by weight)
1	<300	<0,20
2	300–1000	0,20-0,5
3	>1000	>0,5

Analysis of groundwater is preferred to that of clay. In some cases, it may be found that the results of the groundwater and clay analyses lead to different classifications; the more severe classification should be adopted

Table 3.9 British Code of Practice CP 110:1972

	Concentration of sulfates as SO_3		
Class of exposure	In groundwater (ppm)	In 2:1 water-soil	Total in soil (ppm)
1	<300	–	<2000
2	300–1200	–	2000–5000
3	1200–2500	1,9-3,1	5000–10,000
4	2500–5000	3,1-5,6	10,000–20,000
5	>5000	>5,6	>20,000

For classes 3 to 5, if calcium sulfate is predominant, the water–soil extract may give a lower class of exposure than the total SO_3. To convert SO_3 to SO_4, multiply by 1,2

Table 3.10 British Code of Practice CP 8110:1985

	Concentration of sulfates as SO_3		
Class of exposure	In groundwater (g/l)	In 2:1 water-soil extract (g/l)	Total (ppm)
1	<0,3	<1,0	<2000
2	0,3-1,2	1,0-1,9	2000–5000
3	1,2-2,5	1,9-3,1	5000–10,000
4	2,5-5,0	3,1-5,6	10,000–20,000
5	>5,0	>5,6	>20,000

If calcium sulfate is predominant the water–soil extract may give a lower class than total SO_3. To convert SO_3 to SO_4, multiply by 1,2

Table 3.11 British Standard BS 5328:1997 Guide to specifying concrete

Class of exposure	Concentration of sulfates as SO_4			Concentration of magnesium	
		In soil			
	In groundwater (ppm)	By acid extraction (%)	By 2:1 water–soil extract (g/l)	In groundwater (ppm)	In soil (ppm)
1	<400	<0,24	<1,2	–	–
2	400–1400	[a]	1,2-2,3	–	–
3	1500–3000	[a]	2,4-3,7	–	–
4A	3100–6000	[a]	3,8-6,7	≤1000	≤1,2
4B	3100–6000	[a]	3,8-6,7	>1000	>1,2
5A	>6000	[a]	>6,7	≤1000	≤1,2
5B	>6000	[a]	>6,7	>1000	>1,2

Consider both SO_4 and magnesium
Classification on the basis of groundwater is preferred
Higher values are given for the water –soil extract in recognition of the difficulty of obtaining representative samples and of achieving a comparable extraction rate to that from groundwater
[a] Classify on the basis of 2:1 water– soil extract

Table 3.12 British BRE Digest 90 (1968)

Class of exposure	Concentration of sulfates as SO_3		
	In groundwater (ppm)	In 2:1 water-soil extract (g/l)	Total (ppm)
1	<300	–	<2000
2	300–1200	–	2000–5000
3	1200–2500	2,5-5,0	5000–10,000
4	2500–5000	5,0–10,0	10,000–20,000
5	>5000	>10	>20,000

To convert SO_3 to SO_4, multiply by 1,2

Table 3.13 British BRE Digest 174 (1975)

Class of exposure	Concentration of sulfates as SO_3		
	In groundwater (ppm)	In 2:1 water-soil extract (g/l)	Total (ppm)
1	<300	–	<2000
2	300–1200	–	2000–5000
3	1200–2500	1,9-3,1	5000–10,000
4	2500–5000	3,1-5,6	10,000–20,000
5	>5000	>5,6	>20,000

To convert SO_3 to SO_4, multiply by 1,2

Table 3.14 British BRE Digest 250 (1981)

Class of exposure	Concentration of sulfates as SO_3		
	In groundwater (ppm)	In 2:1 water-soil extract (g/l)	Total (ppm)
1	<300	<1,0	<2000
2	300–1200	1,0-1,9	2000–5000
3	1200–2500	1,9-3,1	5000–10,000
4	2500–5000	3,1-5,6	10,000–20,000
5	>5000	>5,6	>20,000

To convert SO_3 to SO_4, multiply by 1,2

Table 3.15 British BRE Digest 363 (1991)

Class of exposure	Concentration of sulfates as SO_4			Concentration of magnesium	
		In soil			
	In groundwater (ppm)	By acid extraction (%)	By 2:1 water–soil extract (g/l)	In groundwater (ppm)	In soil (ppm)
1	<400	<0,24	<1,2	–	–
2	400–1400	[a]	1,2–2,3	–	–
3	1500–3000	[a]	2,4–3,7	–	–
4a	3100–6000	[a]	3,8–6,7	≤1,0	≤1,2
4b	3100–6000	[a]	3,8–6,7	>1,0	>1,2
5a	>6000	[a]	>6,7	≤1,0	≤1,2
5b	>6000	[a]	>6,7	>1,0	>1,2

[a]If >0,24, classify on the basis of 2:1 water–soil extract

Table 3.16 British BRE Special Digest 1 (2002). British Standard BS 8500:2000

Class of exposure	Concentration of sulfates as SO_4		Concentration of magnesium	
	In groundwater (ppm)	In 2:1 water–soil extract (g/l)	In groundwater (ppm)	In 2:1 water–soil extract (g/l)
DS-1	<400	<1200	–	–
DS-2	400–1400	1200–1300	–	–
DS-3	1500–3000	2400–3700	–	–
DS-4a	3100–6000	3800–6700	≤1000	≤1200
DS-4b	3100–6000	3800–6700	>1000	>1200
DS-5a	>6000	>6700	≤1000	≤1200
DS-5b	>6000	>6700	>1000	>1200

Consider both SO_4 and magnesium
Static and mobile water are distinguished
The value of pH of the water is also considered

forms water-soluble calcium compounds, which are then leached away by aqueous solutions. Naturally occurring sulfates of sodium, potassium, calcium, or magnesium are sometimes found in soil or dissolved in groundwater. Sulfates can attack concrete by reacting with hydrated compounds in the hardened cement. These reactions can induce sufficient pressure to disrupt the cement paste, resulting in loss of cohesion and strength. Calcium sulfate attacks calcium aluminate hydrate and forms ettringite. Sodium sulfate reacts with calcium hydroxide and calcium aluminate hydrate forming ettringite and gypsum. Magnesium sulfate attacks in a manner similar to sodium sulfate and forms ettringite, gypsum, and brucite (magnesium hydroxide). Brucite forms primarily on the concrete surface, consumes calcium hydroxide, lowers the pH of the pore solution, and then decomposes the calcium silicate hydrates (PCA, 2002).

Methods of mitigation of the effects of ESA and acid attacks

In reference to structural design considerations, Roads and Traffic Authority (RTA, 2008) advice that, in addition to the normal design process, in the design and construction of any structures in ASS careful consideration of the options available to mitigate the effects of sulfate and acid attack have to be taken into account and be incorporated from the beginning of the concept design stage. Some protective methods such as minimizing soil disturbance, quality of concrete, sacrificial layers, protective coating, addressing site drainage and the like are briefly explained. Any of these measures alone or in combination should be used to achieve durable concrete structures:

- Soil Disturbance: Soil disturbance in ASS should be avoided wherever possible or minimised to limit the extent of oxidization of ASS during construction e.g. driven piles should be selected over cast *in-situ* bored piles; the possibility of extending piles above ground level to the headstocks or having above ground pile caps to avoid excavation should be considered.
- Design Effects: The structural behaviour of any protective measures should be considered during design. If the shaft of a concrete pile is covered by a special casing, liner or coating, account for the reduction in the shaft resistance for that length of pile which is protected. The axial load carrying capacity of friction piles is likely to be greatly reduced compared to mainly end-bearing piles. Where cast-*in-situ* piles have to be installed full depth in highly aggressive ground, use fully sleeved pile shafts on expanded bases of inert aggregate, but the end bearing capacity may be less than when the pile bases are formed with concrete. The use of a sacrificial outer layer for friction piles is not appropriate since the surface of the concrete must remain sound to prevent loss of skin friction or settlement.
- Sacrificial Concrete Layer: Under stagnant soil water conditions, the use of reactive aggregate can mitigate the degree of chemical attack on the concrete e.g. good quality limestone aggregate can neutralise some of the acid that would otherwise attack the cement paste. There are some advantages in having a coarse aggregate which is not completely immune to attack, because the acid concentration can be reduced more rapidly if the acid reacts with the aggregate in addition

to the cement. Since limestone-aggregate concrete is capable of maintaining a reasonably smooth surface during erosion, it can be practicable to design for relatively thick sacrificial layers. When total cover, comprising the design cover plus the sacrificial layer thickness, exceeds 80 mm, supplementary mesh reinforcement in the cover zone is required. The use of limestone aggregate without an increase in concrete cover is disastrous. Therefore it is crucial that when limestone aggregate is specified for ASS an increase in concrete cover as a sacrificial layer is made. The contribution of the increased concrete thickness should be discounted when calculating the design strength of the member.

- Quality concrete for sulfate attack: It is essential that the concrete is dense and well compacted to enable it to resist sulfate attack. Low concrete permeability and choice of cement type is more important than high characteristic strength. Concrete should be fully cured and mature before exposure to sulfates as it has been found that more mature concretes are far more resistant to sulfate attack than immature concretes. In neutral to alkaline environments, some resistance to sulfate attack is obtained when the tricalcium aluminate—C3A—content of the cement is low. Traditional sulfate resisting Portland cements with low C3A have been and are still used in such environments. In acidic environments, the effectiveness of Portland cement of low C3A alone in resisting sulfate attack is questionable. Several investigations has demonstrated that the addition of supplementary cementitious materials to Portland cement improves sulfate resistance in neutral to acidic environments. Several papers report that blended cements give protection against sulfate attack superior to sulfate resisting Portland cement, due to the lower C3A content from the Portland cement replacement, refined pore structure, and reduction of calcium hydroxide available to form gypsum that contributes to the attack. Blended cements are effective only when proportioned correctly. Fly ash blends have good sulfate-resisting properties only when the fly ash content is more than 25% of the cement. Fly ash proportions exceeding 40% are not generally recommended because of placement difficulties and a lack of data on performance. Slag blended cements with more than 65% slag are considered sulfate resisting. The higher the slag content up to a maximum of 75% of the total cement, the higher the resistance to sulfate attack, but such high slag contents should be used only for concrete surfaces either permanently wet or permanently isolated from the air, to guard against the risk of carbonation when exposed to atmospheric carbon dioxide. Silica fume blends of 5–15% of the cement also have good sulfate-resisting characteristics.
- Quality concrete for acid attack: Concrete erosion by acidic water depends more on concrete quality than the type of cement. The lower porosity and permeability of blended cements is generally beneficial. Blended cements incorporating fly ash and/or slag in the proportions discussed above should be used. The use of limestone aggregate in a sacrificial concrete layer is discussed above. For piles installed in an impermeable clay soil, acid or sulfate attack may only penetrate the concrete to such a small extent that only a small thickness of dense sacrificial concrete may remove the need for special cements. In other circumstances where the soil is of higher permeability, additional measures may also be necessary.

- Protective Coatings: Use of coatings for steel members is well established with a wide range of references and standards available. Concrete coatings are covered in many references but not to the extent of those for steel. Concrete coatings should be considered as only one line of defence against an aggressive environment for use with other measures. As a general rule, use of coatings does not permit relaxation of concrete quality and composition discussed above. The selection of a coating for concrete is a complex process and is affected by several factors. These factors include:
  - The type of surface to be coated and its condition;
  - Surface orientation (e.g. horizontal, vertical);
  - Construction methods;
  - The environment and level of aggressiveness;
  - Duration and changes of exposure;
  - Temperature at time of curing of coating;
  - System life required;
  - Maintenance frequency and methods, if feasible.

- Attempting to recommend a general coating system for a specified environment is not appropriate since the environment is not the only factor. Expert advice should be sought on a case by case basis. Coatings should only be applied by trained and experienced applicators.
- There are a number and variety of coating systems which are differentiated by the basis of the coating and the curing process, with the following generic systems recommended for concrete:
  - Chlorinated Rubber;
  - Epoxy: Epoxy coatings generally have satisfactory acid resistance where $pH > 3.0$. They are not suitable where $pH < 3.0$.
  - Vinyl Ester: Vinyl ester coatings are the most appropriate coating for the protection of concrete in acid soil conditions where $pH < 3.0$.
  - Bituminous.

- Other Protection Methods: Protection other than coatings may be required in some circumstances including prolonged exposure to very aggressive environments capable of penetrating any concrete and coating combination. In such circumstances, a higher level of isolation from the environment is required; some methods used are:
  - 10 mm thick permanent mild steel casings with 1.5 mm thick inner liner of heavy duty PVC for piles for extremely high sulfate concentrations e.g. 13,800 to 28,800 ppm;
  - In an environment of 3000 ppm sulfates and high chloride concentrations in the Middle East, PVC sleeves coated with nylon fabric were recommended for pile protection;
  - Packing a layer of limestone around the exposed face of the concrete element for protection of piles in acidic groundwater;

- A sacrificial layer of concrete on piles together with other measures, but this is rarely solely relied on for highly aggressive conditions;
- Local replacement of permeable soil around piles with a layer of impermeable soil.

The following summarises procedures to be followed by designers, specifiers and/or Project Managers in the design, specification and construction of structures in Acid Sulfate Soils. The procedures are grouped in three main stages (RTA, 2008):

1. Investigation:
   (a) Review nature of ASS (actual Acid Sulfate Soils and potential Acid Sulfate Soils)
   (b) Gather data from completed projects with Acid Sulfate Soils
   (c) Obtain existing soil permeability, pH, SO_4 and Mg
   (d) Obtain expert advice on potential pH and SO_4 over lifetime of structure
2. Design and review:
   (a) Determine exposure classification and durability design requirements
   (b) Input into to overall design of structure
   (c) Input into detailed structural concrete design and technical specification, as required
   (d) Specify concrete mix requirements
   (e) Obtain expert advice on type of coatings and specify as appropriate
   (f) Specify other protection methods, as required
3. Construction:
   (a) Monitor compliance with design and specification

At each stage a number of steps is recommended. It is essential that designers, specifiers and project managers communicate at all stages of the project development to deliver adequate, buildable and economical structures. Monitoring and evaluation of design, specification and construction of structures in ASS in a project will not only satisfy that project's requirements but will also benefit other projects in the vicinity.

References

ACI Committee 201. (2001). *Guide to durable concrete* (ACI 201.2R-01, p. 41). American Concrete Institute.

Aïtcin, P.-C. (2004). *High performance concrete*. Taylor & Francis, 621p.

Ammons, J., Shelton, P., & Fulton, B. (1991). Monitoring of chemical properties on a pyritic disposal-area soil after liming. *Soil Science Society of America Journal, 55*, 368–371.

Aslan, A., & Autin, W. (1996). Depositional and pedogenic influences on the environmental geology of Holocene Mississippi River floodplain deposits near Ferriday, Louisiana. *Engineering Geology, 45*, 417–432.

ASTM C1012/C1012M-18, (2018). Standard Test Method for Length Change of Hydraulic-Cement Mortars Exposed to a Sulfate Solution, ASTM International, West Conshohocken, PA.
Baronio, G., & Berra, M. (1984). Concrete deterioration with the formation of thaumasite – Analysis of the causes. *Il Cemento, 3*, 169–184.
Bellmann, F., Erfurt, W., & Ludwig, H.-M. (2012). Field performance of concrete exposed to sulfate and low pH conditions from natural and industrial sources. *Cement and Concrete Composites, 34*(1), 86–93.
Bensted, J. (1988). Thaumasite – A deterioration product of hardened cement structures. *Il Cemento, 85*(1), 3–10.
Berra, M., & Baronio, G. (1987). Thaumasite in deteriorated concretes in the presence of sulfates. In J. M. Scanlon (Ed.), *Proceedings of Katherine and Bryant Mather International Conference on Concrete Durability, Atlanta, 1987. SP-100* (Vol. 2, pp. 2073–2089). American Concrete Institute.
Bickley, J. A., Hemmings, R. T., Hooton, R. D., & Balinski, J. (1993a). Low temperature sulfate attack on arctic concrete structures: A case history. In *Proceedings of 3rd Canadian symposium on cement and concrete* (pp. 57–77). Ottawa.
Bickley, J. A., Hemmings, R. T., Hooton, R. D., & Balinski, J. (1993b). Thaumasite related deterioration of concrete structures. In P. K. Mehta (Ed.), *Proceedings of V Moham Malhotra symposium on concrete technology–past, present and future* (ACI SP-144, pp. 159–175). ACI.
BRE. (2001). *BRE special digest 1: Concrete in aggressive ground.* CRC.
British Cement Association. (1988). *The diagnosis of alkali-silica reaction* (Report of a Working Party). British Cement Association.
Brown, P., & Doerr, A. (2000). Chemical changes in concrete due to the ingress of aggressive species. *Cement and Concrete Research, 30*, 411–418.
Brown, P., & Hooton, R. D. (2002). Ettringite and thaumasite formation in laboratory concretes prepared using sulfate-resisting cements. *Cement & Concrete Composites, 24*, 361–370.
Brueckner, R., Williamson, S. J., & Clark, L. A. (2012). Rate of the thaumasite form of sulfate attack under laboratory conditions. *Cement & Concrete Composites, 34*, 365–369.
Building Research Establishment. (1996). *Digest 363, Sulfate and acid resistance of concrete in the ground.* Building Research Establishment.
Chinchón, S. (2013). *Áridos reactivos en hormigones de presa. Reacción sulfática con formación de thaumasita.* Tesis doctoral, Universidad de Alicante.
Clark, L. A., & BRE in consultation with The Thaumasite Expert Group. (2000a). *Thaumasite Expert Group Report* (Personal communication about occurrence of TSA in concrete of the stairway fan aeration tank at Marbjerg waterworks, Roskilde, from Cowi Organisation, Lyngby, Denmark).
Clark, L. A., & BRE in Consultation with the Thaumasite Expert Group. (2000b). *Thaumasite Expert Group Report. One year review.*
Clark, L. A., & BRE in Consultation with the Thaumasite Expert Group. (2002). *Thaumasite Expert Group Report: Review after three years experience.*
Clifton, J. (1999). Standards for evaluating the susceptibility of cement based materials to external sulfate attack. In *Materials Science of Concrete-Sulfate Attack Mechanisms, Special Volume. Proceedings from Seminar on Sulfate Attack Mechanisms, Quebec, Canada, October 5–6, 1998* (pp. 337–355). American Ceramic Society.
Collepardi, M. (2003). A state-of-the-art review on delayed ettringite attack on concrete. *Cement & Concrete Composites, 25*, 401–407.
Crammond, N. (1985). Thaumasite in failed cement mortars and renders from exposed brickwork. *Cement and Concrete Research, 15*, 1039–1050.
Crammond, N. (2002a). The occurrence of thaumasite in modern construction–a review. *Cement & Concrete Composites., 24*, 393–402.
Crammond, N. (2002b). The thaumasite form of sulfate attack. In *Proceedings of the first international conference on Thaumasite in Cementitious Materials*. BRE.

Crammond, N. J. (2003). The thaumasite form of sulfate attack in the UK. *Cement & Concrete Composites, 25*, 809–818.

Crammond, N. J., & Dunster, A. D. (1997). *Avoiding the deterioration of cement-based building materials, Lessons from case studies 1* (BRE Lab. Report BR 324). Construction Research Communications (CRC).

Crammond, N. J., & Halliwell, M. A. (1995). The thaumasite form of sulfate attack in concretes containing a source of carbonate ions – A microstructural overview. *ACI Special Publications, 154*, 357–380.

Crammond, N. J., Longworth, T. I., & Sibbick, R. G. (1999). *Report on occurrence of thaumasite form of sulfate attack to concrete foundation at Abbeymead, Gloucester* (Report No. 80414). BRE.

Dehwah, H. (2007). Effect of sulfate concentration and associated cation type on concrete deterioration and morphological changes in cement hydrates. *Construction Building Materials, 21*, 29–39.

Deloye, F.-X., Louarn, N., & Loos, G. (1988). Examples of masonry analysis. The case of the Puberg tunnel. *Bull de Liaison Lab des Ponts et Chaussees, 163*, 17–24.

Dent, D. (1986). *Acid sulfate soils: A baseline for research and development*. International Institute for Land Reclamation and Improvement/ILRI.

Drimalas, T., Clement, J. C., Folliard, K. J., Dhole, R., & Thomas, M. D. A. (2011). *Laboratory and field evaluations of external sulfate attack in concrete. Sulfate resistance of concrete exposed to external sulfate attack* (Report FHWA/TX-11/0-4889-1). Texas Department of Transportation.

Duval, R., & Hornain, H. (1992). Chapitre 9: La durabilité du béton vis-à-vis des eaux agressives. In J. Baron & J.-P. Ollivier (Eds.), *La durabilité des bétons*. Presses de l'ENPC.

Eriksen, K. (2003). Thaumasite attack on concrete at Marbjerg Waterworks. *Cement & Concrete Composites, 25*, 1147–1150.

Eriksen, K., & Buhr, B. (2013). Microscopy of Faroe bridges concrete from below sea level. In *14th Euroseminar on Microscopy Applied to Building Materials Helsingor*. Denmark.

Erlin, B., & Stark, D. C. (1965). Identification and occurrence of thaumasite in concrete. Symposium on the effects of aggressive fluids in concrete. *Highway Research Record, 113*, 108–113.

FAO/EC/ISRIC. (2003). *World soil resources*. FAO/EC/ISRIC.

FAO-UNESCO. (2007). *Soil map of the world*. FAO-UNESCO.

Fischer, G., Nachtergaele, F., Prieler, S., van Velthuizen, H. T., Verelst, L., & Wiberg, D. (2008). *Global agro-ecological zones assessment for agriculture (GAEZ 2008)*. IIASA/FAO.

Freyburg, E., & Berninger, A. M. (2003). Field experiences in concrete deterioration by thaumasite formation: Possibilities and problems in thaumasite analysis. *Cement and Concrete Composites, 25*(8), 1105–1110.

Gaze, M. E., & Crammond, N. J. (2000). The formation of thaumasite in a cement: Lime:Sand mortar exposed to cold magnesium and potassium sulfate solutions. *Cement & Concrete Composites, 22*, 209–222.

Geoffrey, G. E., & Alexander, M. G. (2011). *Alkali-aggregate reaction and structural damage in concrete*. CRC Press, Taylor and Francis Group.

Gollop, R. S., & Taylor, H. F. W. (1992). Microstructural and microanalytical studies of sulfate attack. I. Ordinary Portland cement paste. *Cement and Concrete Research, 22*, 1027–1038.

Hagelia, P., et al. (2001). Thaumasite and subsequent secondary calcite deposition in sprayed concretes in contact with sulfide bearing Alum Shale. In P. Hagelia, et al. (Eds.), *Thaumasite and secondary calcite in some Norwegian concretes*.

Hagelia, P., Sibbick, R. G., Crammond, N. J., & Larsen, C. K. (2003). Thaumasite and secondary calcite in some Norwegian concrete. *Cement and Concrete Composites, 25*(8), 1131–1140.

Hartshorn, S. A., Sharp, J. H., & Swamy, R. N. (1999). Thaumasite formation in Portland-limestone cement pastes. *Cement and Concrete Research, 29*, 1331–1340.

Haynes, H., O'Neill, R., & Mehta, P. K. (1996). Concrete deterioration from physical attack by salts. *Concrete International, 18*(1), 63–68.

Higgins, D., & Crammond, N. (2002, June). *Resistance of concrete containing ggbs to the thaumasite form of sulfate attack*. Paper presented at the first international conference on thaumasite in cementitious materials, Watford, UK.

Hobbs, D. W. (1988). *Alkali-silica reaction in concrete*. Thomas Telford.

Hobbs, D. W. (2003). Thaumasite sulfate attack in field and laboratory concretes: Implications for specifications. *Cement & Concrete Composites, 25*, 1195–1202.

Hobbs, D. W., & Taylor, M. G. (2000). Nature of the thaumasite sulfate attack mechanism in field concrete. *Cement and Concrete Research, 30*, 529–533.

Hooton, R. D., Ramezanianpour, A., & Ahani, R. (2013). *Sulfate resistance of concretes containing Portland-limestone cement & limestone fillers*. Presentation at the TC 251-SRT meeting, Paris.

Hu, M. Y., Tang, M. S., & Long, F. M. (2004). Thaumasite form of sulfate attack in the concrete of Yongan Dam. *Concrete, 11*, 5–7. (in Chinese).

Hu, M. Y., Long, F. M., & Tang, M. S. (2006). The thaumasite form of sulfate attack in concrete of Yongan Dam. *Cement and Concrete Research, 36*, 2006–2008.

Hustnes, H., & Rodum, E. (2006). Case studies of thaumasite formation. In *Seventh CANMET/ACI international conference on Durability of Concrete*. ACI.

Irassar, E. F., Di Maio, A., & Batic, O. R. (1995). Sulfate attack on concrete with mineral admixtures. *Cement and Concrete Research, 26*(1), 113–123.

Jacobsen, S., Lindgard, L., & Fritz, L. (2008). Frost dilation measurements on concrete cores from dam with ASR. In *Proceedings on international congress on Alkali-Aggregate Reaction* (pp. 131–142). Thondheim, Norway.

Jones, J. C. (2000). Influences of soil acidity levels on vegetative reclamation and wildlife habitat on rights-of-ways transecting drastically disturbed lands. In J. Goodrich-Mahoney et al. (Eds.), *Proceedings, seventh international symposium on Environmental Concerns in Rights-of-Way Management, 9–13 Sept*. Elsevier Science.

Khatri, R. P., & Sirivivatnanon, V. (1997). Role of permeability in sulfate attack. *Cement and Concrete Research, 27*, 1179–1189.

Khomwan, N., Puktrap-Anan, P., & Srisura, R. (2009). Deterioration of RC structure after long term exposure. *Proceeding of the Sixth Regional Symposium Infrastructure Development, 26*(10), 1553–1996.

Leemann, A., & Loser, R. (2011). Analysis of concrete in a vertical ventilation shaft exposed to sulfate-containing groundwater for 45 years. *Cement & Concrete Composites, 33*, 74–83.

Lei, M., Peng, L., Shi, C., & Wang, S. (2013). Experimental study on the damage mechanism of tunnel structure suffering from sulfate attack. *Tunnelling and Underground Space Technology, 36*, 5–13.

Lipus, K., & Punkte, S. (2003). Sulfatwiderstand unterschiedlich zusammengesetzter Betone, Tl.1 [Sulfate resistance of concretes with different compositions (part 1)]. *Beton Dusseldorf, 53*(Part 2), 97–101.

Long, G. C., Xie, Y. J., Deng, D. H., & Li, X. K. (2011). Deterioration of concrete in railway tunnel suffering from sulfate attack. *Journal of Central South University of Technology, 18*, 881–888.

Lossing, F. A. (1966). Sulfate attack on concrete pavements in Mississippi. In *Symposium on effects of aggressive fluids on concrete* (Vol. 113, pp. 88–102). Highway Research Record.

Loudon, N. (2003). A review of the experience of thaumasite sulfate attack by the UK highways agency. *Cement & Concrete Composites, 25*(8), 1051–1058.

Ma, B. G., Gao, X. J., Byars, E. A., & Zhou, Q. Z. (2006). Thaumasite formation in a tunnel of Bapanxia Dam in Western China. *Cement and Concrete Research, 36*(4), 716–722.

Ma, K. L., Long, G. C., & Xie, Y. J. (2012). Railway tunnel concrete lining damaged by formation of gypsum, thaumasite and sulfate crystallization products in southwest of China. *Journal of Central South University, 19*, 2340–2347.

Mehta, P. K. (2000). Sulfate attack on concrete: Separating myths from reality. *Concrete International, 22*(8), 57–61.

Mehta, P. K., Schiessl, P., & Raupach, M. (1992). Performance and durability of concrete systems. In *Proceedings of the 9th international congress on the Chemistry of Cement* (Vol. 1, pp. 571–662). New Delhi, India.

Menéndez, E. (2002). Cracking and sulfate attack in field concrete in Spain. In *RILEM workshop on Internal Sulfate Attack and Delayed Ettringite Formation* (pp. 127–138). Villars, Switzerland.

Menéndez, E. (2010). *Analysis of concrete structures affected by alkali silica reaction, sulfate attack and frost-drawing cycles (in Spanish)*. IECA.

Menéndez, E., Gil, A., Martín, A.B., Aldea, B., & Marquez, C. (2011). Analysis of concrete dams affected by internal expansive processes. In *Differences and similarities between internal sulfate attack and alkali-aggregate reaction (in Spanish), Advance in Security and Durability, Fundetel* (pp. 283–306).

Mittermayr, F., Baldermann, A., Klammer, D., Leis, A., Tritthart, J., & Dietzel, M. (2013). Sulfate attack – Reaction mechanisms revealed by a multi proxy approach. In *Civil and Environmental Research, Vol. 5. Special issue for International Congress on Materials & Structural Stability* (pp. 27–30). Rabat, Morocco.

Mittermayr, F., Deméný, A., Thaller, D., Leis, A., Czuppon, G., Dietzel, M. (2015). Concrete corrosion in tunnels: A stable O and H isotope study of TSA mechanism. In *11th applied isotope geochemistry conference, AIG-11 BRGM. Procedia Earth and Planetary Science* (Vol. 13, pp. 248–251).

Moum, J., & Rosenqvist, I. T. (1959). Sulfate attack on concrete in the Oslo Region. *Journal of the American Concrete Institute, 31*(3), 257–264.

Nehdi, M. L., Suleiman, A. R., & Soliman, A. M. (2014). Investigation of concrete exposed to dual sulfate attack. *Cement and Concrete Research, 64*, 42–53.

Neville, A. (2004). The confused world of sulfate attack on concrete, review. *Cement and Concrete Research, 34*, 1275–1296.

NSW Acid Sulfate Soils Management Advisory Committee. (1998). *Acid sulfate soils assessment guidelines*. NSW.

Oberholster, R. E., Van Aardt, J. H. P., & Brandt, M. P. (1983). Durability of cementitious systems. In P. Barnes (Ed.), *Structure and performance of cements* (pp. 365–413). Applied Science Publishers.

Oberholster, R. E., Maree, H., & Brand, J. H. B. (1992). Cracked prestressed concrete railway sleepers: Alkali-silica reaction or delayed Ettringite formation. In *Proceedings of the 9th international conference of Alkali-Aggregate Reaction in Concrete* (pp. 739–749). London, UK.

Ostnor, T. A., Justnes, H., & Franke, W. (2014, December). A condition survey of concrete elements with corrosion inhibitor exposed 12 years for sea water in tidal zone. In *2nd International congress on Durability of Concrete as paper* (No. 64, pp. 4–6). New Delhi, India.

Park, K. (1996, November). Historic Manitoba aqueduct undergoes extensive repairs. *Heavy Construct News*, pp. 16–17.

Pauri, M., & Collepardi, M. (1989). Thermo-hygrometrical stability of thaumasite and ettringite. *Il Cemento, 86*, 177–184.

PCA. (2002). *Types and causes of concrete deterioration* (PCA R&D Serial No. 2617). PCA.

Rahman, M., & Bassuoni, M. (2014). Thaumasite sulfate attack on concrete: Mechanisms, influential factors and mitigation. *Construction and Building Materials, 73*, 652–662.

Rogers, C. A., Thomas, M. D. A., & Lohse, H. (1997). Thaumasite in Manitoba and Ontario, Canada. In *19th international conference on Cement Microscopy* (pp. 306–319). Cinncinnati.

Romer, M. (2003). Steam locomotive soot and the formation of thaumasite in shotcrete. *Cement & Concrete Composites, 25*(8), 1173–1176.

RTA Bridge Engineering. (2008). *Acid sulfate soils. Concrete structures-advice for design and construction* (Review report, Edition 3). RMS.

Sahu, S., Badger, S., & Thaulow, N. (2002). Evidence of thaumasite formation in Southern California concrete. *Cement and Concrete Composites, 24*, 379–384.

Sahu, S., Badger, S., & Thaulow, N. (2003). Mechanism of thaumasite formation in concrete slabs on grade in Southern California. *Cement & Concrete Composites, 25*, 889–897.

Santhanam, M., Cohen, M., & Olek, J. (2001). Sulfate attack research – Whither now? *Cement and Concrete Research, 31*, 845–851.
Sarkar, S. L., & Little, D. N. (1998). Microstructural investigation of severe distress in a crushed concrete base. *Cement and Concrete Research, 28*(3), 401–410.
Scherer, G. W. (2004). Stress from crystallization of salt. *Cement and Concrete Research, 34*(9), 1613–1624.
Schwander, H. W., & Stern, W. B. (1988). Experimental production of transformation products on hardened cement paste surfaces. *Zement-Kalk-Gips, 1*, 81–84. [English translation].
Scrivener, K., & Skalny, J. (2002). *Internal sulfate attack and delayed ettringite formation*. RILEM Publications S.A.R.I.
Shayan, A., & Quick, G. W. (1992). Microscopic features of cracked and uncracked concrete railway sleepers. *ACI Materials Journal, 89*, 348–361.
Shi, C., Wang, D., & Behnood, A. (2012). Review of thaumasite sulfate attack on cement mortar and concrete. *Journal of Materials in Civil Engineering, 24*(12), 1450–1460.
Sibbick, R. G., & Crammond, N. J. (1999) Two case studies into the development of the thaumasite form of sulfate attack in hardened concretes. In: *Proceedings of the 7th Euroseminar on Microscopy Applied to Building Materials*. Delft.
Sibbick, T., Fenn, D., & Crammond, N. (2003). The occurrence of thaumasite as a product of seawater attack. *Cement & Concrete Composites, 25*, 1059–1066.
Sims, I., & Huntley, S. A. (2004). The thaumasite form of sulfate attack-breaking the rules. *Cement & Concrete Composites, 26*(7), 837–844.
Skalny, J., Marchand, J., & Odler, I. (2002). *Sulfate attack on concrete*. Taylor and Francis Group.
Stark, D. C. (1989). *Durability of concrete in sulfate-rich soils* (Research and Development Bulletin, Vol. RD097). Portland Cement Association.
Stark, D. C. (2003). Occurrence of thaumasite in deteriorated concrete. *Cement and Concrete Composites, 25*, 1119–1121.
Stutzman, P. E. (1999). *Deterioration of Iowa highway concrete pavements. A petrographic Study*. NIST.
Suput, J. S., Mladenovic, A., Cernilogar, L., & Olensek, V. (2003). Deterioration of mortar caused by the formation of thaumasite on the limestone cladding of some Slovenian railway tunnels. *Cement & Concrete Composites, 25*(8), 1141–1145.
Tepponen, P., & Eriksson, B. E. (1987). Damages in concrete railway sleepers in Finland. *Nordic Concrete Research, 6*, 199–209.
Thaumasite Expert Group. (1999). *The thaumasite form of sulfate attack: Risks, diagnosis, remedial works and guidance on new construction*. DETR.
Thomas, M., & Ramlocham, T. (2002). Field cases of delayed ettringite formation. In *RILEM workshop on Internal Sulfate Attack and Delayed Ettringite Formation* (pp. 85–97). Villars, Switzerland.
Thomas, M. D. A., Rogers, C. A., & Bleszynski, R. F. (2003). Occurrences of thaumasite in laboratory and field concrete. *Cement and Concrete Composites, 25*(8), 1045–1050.
Thomas, M., Folliard, K., Drimalas, T., & Ramlochan, T. (2008). Diagnosing delayed ettringite formation in concrete structures. *Cement and Concrete Research, 38*, 841–847.
van Aardt, J. H. P., & Visser, S. (1975). Thaumasite formation: A cause of deterioration of Portland cement and related substances in the presence of sulfates. *Cement and Concrete Research, 5*, 225–232.
Van Hees, R. P. J., Wijffels, T. J., & van der Klugt, L. J. A. R. (2003). Thaumasite swelling in historic mortars: Field observations and laboratory research. *Cement and Concrete Composites, 25*, 1165–1171.
Weiss. (2010). *Controlling sulfate attack in Mississippi Department of Transportation Structures* (ERDC/GSL TR-10-31). U.S. Army Engineer Research and Development Center Geotechnical and Structures Laboratory.
West, G. (1996). *Alkali aggregate reaction in concrete roads and bridges*. Thomas Telford.

Whittaker, M., & Black, L. (2015). Current knowledge of external sulfate attack. *Advances in Cement Research, 27*(9), 532–545.

Yoshida, N., Matsunami, Y., Nagayama, M., & Sakai, E. (2010). Salt weathering in residential concrete foundation exposed to sulfate-bearing ground. *Journal of Advanced Concrete Technology, 8*(2), 121–134.

Zarzar, F., & Almeida de Oliveira, R. (2011). Deterioration of foundation elements of buildings, due to sulfates attack. In: *XII DBMC*. Porto, Portugal.

Chapter 4
Review of Existing Lab Test Methods for External Sulfate Attacks and Recommendations for an Improved Testing Methodology

Véronique Baroghel-Bouny, Myriam Carcassès, Franck Cassagnabère, Nele De Belie, Chiara Ferraris, Fredrik P. Glasser, Li-ping Guo, Andreas Leemann, Klaus Lipus, Ahmed Loukili, Mathias Maes, Esperanza Menéndez Méndez, Samuel Meulenyzer, Florian Mittermayr, Othman Omikrine-Metalssi, Antonio Princigallo, Emmanuel Roziere, Manu Santhanam, Karen Scrivener, Anya Vollpracht, and Kai Tai Wan

---

V. Baroghel-Bouny (✉) · O. Omikrine-Metalssi
University Gustave Eiffel, Marne-la-Vallée, France
e-mail: veronique.baroghel-bouny@univ-eiffel.fr

M. Carcassès · F. Cassagnabère
LMDC - INSA/UPS Dept Genie Civil, Toulouse, France

N. De Belie
Ghent University, Ghent, Belgium

C. Ferraris
NIST, Gaithersburg, MD, USA

F. P. Glasser
University of Aberdeen, Aberdeen, UK

L.-p. Guo
School of Materials Science and Engineering, Southeast University, Nanjing, China

A. Leemann
EMPA, Dübendorf, Switzerland

K. Lipus
VDZ, Düsseldorf, Germany

A. Loukili · E. Roziere
GeM – Ecole Centrale de Nantes, Nantes, France

M. Maes
SANACON (former Ghent University), Merelbeke-Melle, Belgium

E. Menéndez Méndez
IETcc-CSIC, Madrid, Spain

V. Baroghel-Bouny et al. (eds.), *External Sulfate Attacks - Mechanisms, Field Behavior and Lab Test Methods*, RILEM State-of-the-Art Reports 48,
https://doi.org/10.1007/978-3-032-19402-2_4

S. Meulenyzer
Lafarge-Holcim R&D, Lyon, France

F. Mittermayr
University of Innsbruck (former TU Graz), Innsbruck, Austria

A. Princigallo
(Italcementi) Heidelberg Materials, Bergamo, Italy

M. Santhanam
IIT Madras, Chennai, India

K. Scrivener
EPFL, Lausanne, Switzerland

A. Vollpracht
RWTH Aachen, Aachen, Germany

K. T. Wan
Brunel University of London, Uxbridge, UK

Abstract Little is known about external sulfate attacks (ESA) when dealing with new concrete mixtures such as low-CO_2 binders. Resistance to external sulfate attacks is therefore a key challenge for the development of new mixtures and for structural durability. However, tools dedicated to accurately quantify and predict the phenomena are missing and more research is needed on this topic to better understand and capture the degradation occurring in the field. In particular, the existing ESA test procedures are not always appropriate. A more inclusive assessment of surface damage, including both expansive (scaling) and softening or disintegration types of destruction, is required.

After the review of the ESA mechanisms and the report on field aspects and observed cases, this chapter presents a review and analysis of the current standards applied in various countries (e.g., ASTM, CSA, SIA, etc.) and of some other existing lab test methods (in accelerated conditions). The various parameters influencing the test results are in particular reviewed. Then, a focus on recent developments and promising research is introduced. The methods reviewed in this chapter are applied to estimate if the tested cement can be regarded as a sulfate-resisting (SR) cement or are dedicated to the assessment (in the lab) of the resistance of concrete exposed to ESA.

The main features of each test method are described in this chapter, such as:

- type of mixtures used (cement paste, mortar, or concrete),
- type of binders used,
- limitations on cement types if any,
- sample dimensions,
- preconditioning procedure,
- test conditions: type of (sulfate) solution, sulfate concentration, temperature, pH constant or not, wetting-drying cycles, etc.
- parameters measured: expansion, visual changes, mass loss, loss of tensile strength, loss of compressive strength, dynamic modulus of elasticity, etc.

In the review and critical analysis of the available test methods, the following important aspects have been considered:

- test methods on both mortar and concrete (and cement paste) specimens,
- relevant parameters for evaluation (and ranking),
- with regard to existing test methods:
  - analysis of agreements/disagreements,
  - investigation of the improvements needed,
- with regard to new test methods:
  - impact of acceleration of the processes,
  - innovative methods of acceleration of the processes.

In addition, recommendations for a new testing methodology are proposed, as a basis for future European standardization on this topic along with dedicated RRTs (e.g., within the framework of CEN/TC51-TC104/WG12/TG1 "Sulfate resistance to external sulfate attack"). Moreover, the questions still open, perspectives and further steps needed, as well as the possible adaptation of tests/models to low-CO_2 binders, are briefly discussed at the end of the chapter.

Keywords (Accelerated) lab test · Mortar · Concrete · Ettringite · Standard · ESA · Recommendations · Test method · Thaumasite

4.1 Introduction: Background and Needs for Sulfate Resistance Testing (in Lab)

A literature review of the situation at the starting date of the Technical Committee has highlighted the following aspects.

4.1.1 Background: Concrete Structures

Nowadays, there is a need for durable and long-lasting structures, even in very aggressive environments as in the case of energy supply including power plants, biogas, ….

Unfortunately, External Sulfate Attack (ESA), which is a **complex** form of deterioration (Neville, 2004; Whittaker & Black, 2015), has damaged a significant number of existing concrete structures and foundations in many countries throughout the world and has thus affected their long-term durability (see, e.g., Feng et al., 2015; Ikumi et al., 2016; Maes & De Belie, 2014; Menéndez et al., 2020b; Irassar et al., 2003; Yang & Guo, 2014; Yoshida, 2019).

ESA is hence often regarded as the second major cause of structure degradation (after reinforcement corrosion) in the world. Among the cases identified of damage by ESA, most of them are observed in:

- foundations in contact with sulfate-containing water or with sulfate-rich soil (e.g., containing gypsum),
- sewer, culvert, buildings for farm use,
- pressure pipes, elements in direct contact with seawater,
- tunnel lining, etc.

In general, the damage can be assigned to ingress of sulfate ions present in surrounding sulfate-rich environments (seawater or underground). However, various forms, with regard to the reactions and detrimental products, of ESA exist, depending on the environmental conditions.

ESA can be characterized by:

Ettringite or/and **gypsum** formation. In this case, the reaction between sulfates and portlandite (after its dissolution) induces gypsum formation:

$$Ca(OH)_2 + SO_4^{2-} + H_2O \rightarrow CaSO_4 \cdot 2H_2O + 2\,OH^-$$

which can react later with anhydrous or hydrated aluminates (e.g., AFm) to produce ettringite (Matschei et al., 2007):

$$C_3A \text{ or } C-A-H + CaSO_4 \cdot 2H_2O + H_2O \rightarrow C_3A \cdot 3CaSO_4 \cdot 32H_2O$$

It is worth mentioning here that high sulfate concentration will favour gypsum formation.

and/or **thaumasite** formation:

$$C-S-H + CaSO_4 \cdot 2H_2O + H_2O + CaCO_3 \rightarrow CaSiO_3 \cdot CaSO_4 \cdot CaCO_3 \cdot 15H_2O$$

Thaumasite can be found in deep foundations. Thaumasite formation is favoured at low temperature (typically $T \leq 5$ °C, see Sect. 4.2.2.3) and in the presence of carbonates (e.g., carbonate-rich aggregates) (Crammond & Halliwell, 1995). However, in field conditions, it can be formed in various types of climates (Torres et al., 2011). The *thaumasite form of sulfate attack* can be more harmful than formation of ettringite and gypsum only, and can destroy the cementing properties of the C-S-H matrix (Crammond, 2002). However, it is worth noting that it does not depend on the C_3A content of the cement (Schmidt et al., 2008). According to (Menéndez et al., 2020a), ettringite, thaumasite and gypsum can be all found in mortar bars exposed to Na_2SO_4 solutions at $T = 5$ °C. This indicates that the various possible modes of deterioration by ESA are not always occurring independently.

ESA yields expansion/cracking of the material, mass or mechanical strength loss, etc. (Santhanam et al., 2001; Chabrelie, 2010; Menéndez et al., 2019) (see Fig. 4.1). At the scale of the structure, degradation is observed such as spalling/bursting of concrete cover, decrease in mechanical properties, alteration of reinforcement-concrete bonding (Jabbour, 2019), etc.

Damages in this case are due to both (coupled):

a) field sample after 19 y under partial immersion in sulfate enriched ground (Chabrelie, 2010)

b) thaumasite form of ESA (Barker and Hobbs, 1999)

Fig. 4.1 Examples of ESA-induced degradations. (**a**) field sample after 19 years under partial immersion in sulfate enriched ground (Chabrelie, 2010). (**b**) thaumasite form of ESA (Barker & Hobbs, 1999)

- transport processes, implying mineralogical changes and gradients between the surface of the concrete body and the core regions by an uptake of sulfates, often combined with leaching,
- chemical reactions, yielding precipitation of secondary sulfate phases (Schmidt et al., 2009; Lothenbach et al., 2010). Note that ettringite formation does not systematically yield damage.

A theory has been proposed to explain these observations: the expansive forces are due to the crystallization pressure of ettringite (Ping & Beaudoin, 1992a, b; Scherer, 2004; Flatt & Scherer, 2008). Two conditions are required for the occurrence of crystallization pressure, which yields stresses and hindered expansions (see also Yu et al., 2013; Feng et al., 2015):

- confined space, where a solid product is formed and grown,
- activity product of the reactants in the pore solution greater than solubility product under atmospheric pressure (supersaturation of the pore solution).

Sulfate attack by physical mechanisms (so-called *physical sulfate attack* (Neville, 2004), see Chap. 2, and Sects. 4.2.3, 4.8.3 and 4.8.4) is also possible. In this case, significant damage is observed (Zhutovsky & Hooton, 2017; Menéndez et al., 2019), which can be attributed to the **crystallisation of sodium sulfate salts**, according to the theory of the crystallization pressure (Lafuma, 1929; Scherer, 2004; Flatt & Scherer, 2008) (see Sects. 4.2.3, 4.8.3 and 4.8.4). *Physical sulfate attack* occurs for example in the case of strong and long periods of drying, which take place in arid zones such as Western USA, Canada, Australia, Middle East of China, and south of Spain, and also in foundations, slabs partially above ground and structures in tidal zones, where both capillary absorption and evaporation of sulfate-bearing water can occur (Dhole et al., 2019). Such sulfate attacks have also been observed in residential concrete foundations, in various areas in Japan (Yoshida, 2019). The mentioned transport processes yield an increase in the sulfate concentration in concrete. In addition, repeated cycles, initiated by temperature and RH changes, enhances the phenomenon. The pore solution may then become supersaturated in mirabilite. Crystal efflorescence at the concrete surface, or crystal growth in pores below the surface (in the near-surface zone pores), will then occur (Henrique de Bem & Medeiros, 2020). In this last case, high crystallization pressure may occur even in large pores. These "subflorescences", which form at the drying front, may be critical if the crystallization pressure exceeds the tensile strength of the material (Scherer, 2004).

ESA is not a new topic; it has been studied worldwide on various cementitious materials for a very long time. In addition, it has been recognized as a cause of concrete distress for many years. However, ESA degradation remains a controversial, confusing, and complex topic. In particular, the consequences of ESA on the material microstructure and durability-related properties are not fully elucidated. The complexity is enhanced by the fact that the underlying mechanisms can vary with the exposure (and observation) time. Consequently, no "universally" adopted test method dedicated to ESA is available, despite significant efforts on the development of lab tests for a long time and continuous improvement of the applied test procedures.

4.1.2 Background: European Standardization

Initially, the recommendations/specifications established to insure durability with regard to ESA were based on the composition of cements (e.g., sulfate resisting cements, see SR qualification in EN 197-1:2012) and concretes. Such recommendations/specifications could provide quite acceptable durability. However, difficulties have arisen, owing to national differences in the types of cements which were recognized as SR and owing to the development of new types of cements/binders (in particular low-CO_2 binders).

In addition, accelerated tests in lab were proposed to evaluate the performance of cements in sulfate-rich environments (e.g., ASTM standard test methods). These tests are based on full immersion of mortar specimens in sulfate solutions and on measurement of the length change as a function of time (see, e.g., Figure 4.2). These tests last in general at least 6 months, except for very rapid tests such as those (14 days) described in the ASTM C452-15 test procedure (2015).

Experimental work has been carried out during the 1990s within the framework of the task group CEN/TC51/TC104/WG12/TG1 "*Sulfate resistance testing*" (European standardization). However, after a decade of work, the following conclusion was drawn:

- most of the tests were regarded as "very long",
- disagreement on the performance of cements containing SCMs,
- poor reproducibility of the applied accelerated lab tests and important discrepancies between performances in field and in lab,

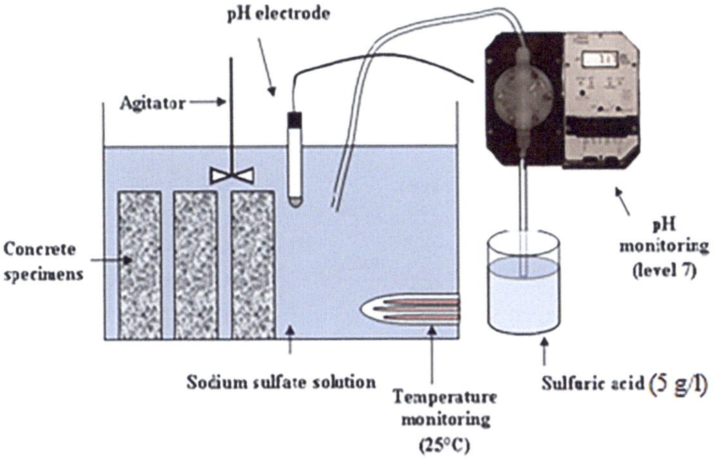

Fig. 4.2 Example of setup for accelerated ESA tests in lab (Messad, 2009; Linger et al., 2014)

- the test methods studied could not be used to predict concrete performance in field.

In particular, the test conditions were considered as very far from field conditions. For example, these tests often use concentrations of sulfates of higher order of magnitude than those found in natural environments. This does result in an acceleration of the degradations, but it has been shown that the degradation mechanism is also changed. It can therefore be desirable to use less concentrated solutions to obtain more "representative" tests in lab.

The conclusion of this work was included in the **CEN/TR 15697 report published in 2008** by the task group (CEN/TR 15697, 2008). This CEN report includes a state-of-the-art concerning sulfate resistance testing, highlighting key parameters and giving suggestions for future work. This CEN report stands as a key step in the development of a test method dedicated to ESA thanks to the review of the great number of performance tests used to assess the resistance of cements to ESA and of the round robin tests carried out on mortar specimens on these bases within the CEN standardization framework. Summary tables reported in it are particularly useful. In addition, 250 papers and reports, published between 1970 and 2006, were reviewed for the drafting of this report. This report clearly pointed out the weaknesses of the available tests and in particular their poor reproducibility.

Several RRT were carried out within the framework of CEN/TC51/TC104/WG12/TG1. The modifications to the method included in the fourth RRT improved reproducibility. The test method ("*flat prism method*") is based on:

- Na_2SO_4 solution (16 g/L SO_4^{2-}),
- T = 20 °C,
- full immersion,
- solution replacement every 28 days,

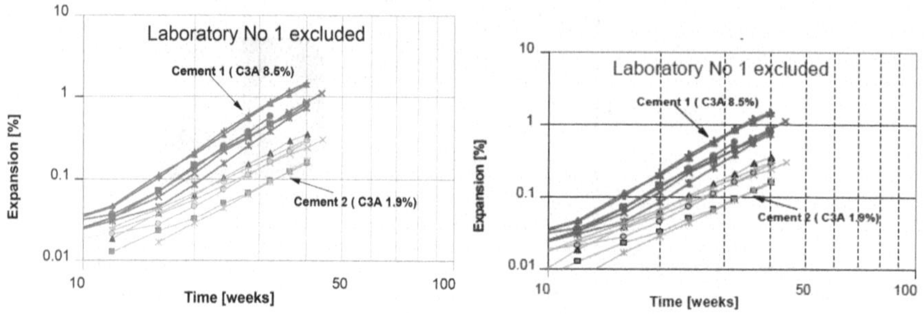

a) Mix A (W/C=0.50, 60 jolts) b) Mix B (W/C=0.60, 60 jolts)

Fig. 4.3 Results from CEN/TC51-TC104/WG12/TG1 fourth testing programme (CEN/TR 15697, 2008). (**a**) Mix A (W/C = 0.50, 60 jolts) (**b**) Mix B (W/C = 0.60, 60 jolts)

- 20 × 20 × 160 mm³ mortar prisms,
- measurement of expansion.

Figure 4.3 illustrates the discrimination achieved between a moderate [C_3A] CEM I cement and a low [C_3A] CEM I SR cement. Whilst within laboratory repeatability was satisfactory at ~10%, the between laboratory reproducibility was poor and in the range 20–50%. If a test procedure is to be used for the acceptance or rejection of a cement or related binder, the coefficient of reproducibility should rather be ≤15%. Note that the coefficient of reproducibility of the EN 196-1: 2016 compressive-strength test is 6% at 28 days.

Consequently, no test procedure was revealed as suitable to be a performance test.

4.1.3 Background: Other Working Groups and Projects

Within the framework of the **NANOCEM European Consortium** (Nanocem, 2004–2021) started in 2004 and associated **Marie Curie Research Training Network (MC-ITN)** on "*Fundamental understanding of cementitious materials for improved, chemical, physical and aesthetic performance*" (leader: Prof. K. Scrivener, EPFL, Lausanne), a comprehensive research work has been conducted over the 5-year 2006–2011 period. The programme included 10 PhD students and also post-doctoral students who worked more or less on the topic of ESA. More precisely, some projects were directly related to the deterioration of cementitious materials by sulfate ions and to the understanding and characterization of sulfate resistance. The scientific work was mainly based on lab experiments, but some of the projects were also oriented on modelling or field conditions with the aim to relate the behaviour in sulfate-bearing environments and phase assemblage (see, e.g., Chabrelie, 2010). The projects were dedicated to the understanding of the basic mechanisms related to sulfate degradations (see, e.g., Yu et al., 2012; Kunther et al., 2013), thanks to analysis at the nanoscale, and the use of new and very accurate methods (e.g., nanoindentation, molecular dynamics modelling, etc.). Most of the projects focused on cements with SCMs (see, e.g., Fernàndez-Altable et al., 2011). The influence of the material composition (W/C, GGBS content, presence of gypsum, …), as well as the effect of the test conditions (type of sample, paste-mortar-concrete, composition and pH of the contact solution, static/dynamic, …), on the observed performance (length change criterion) has been analyzed and the key role of the (confined) precipitation of ettringite in C-S-H gel pores was in particular pointed out on the expansion (Yu et al., 2013). A lot of useful results were published from these research works.

In addition, the **RILEM Technical Committee TC 211-PAE** "*Performance of cement-based materials in aggressive aqueous environments*" (Chair: Prof M. Alexander, Univ. of Cape Town, South Africa) published a comprehensive State-of-the-Art Report (STAR), which included in particular a review on the mechanisms of degradation associated with sulfate attack of concrete in its Part I (see also Glasser, 2009) and a review of sulfate test methods in its Part III (Chap. 10). This STAR (RILEM, 2013) provided thus a basic reference for the present report.

Furthermore, in the STAR entitled "Properties of fresh and hardened concrete containing supplementary cementitious materials" (RILEM, 2018), prepared by Working Group 4 of **RILEM Technical Committee 238-SCM** "*Hydration and microstructure of concrete with supplementary cementitious materials*" (Chair: Prof. N. De Belie, Univ. Ghent, Belgium), the effect of each SCM investigated (GGBS, FA, SF, limestone powder, metakaolin, natural pozzolans, …, as well as ternary blends) on the behavior of concrete with regard to ESA (among other properties/processes) has been addressed. Hence, this STAR contains useful information with regard to the specificities of SCMs, in view of sulfate resistance testing.

4.1.4 *Needs for Sulfate Resistance Testing (in Lab)*

However, despite the large amount of publications and research works about the ESA process, there still exist some open questions, in particular on the underlying **mechanisms**. The complexity begins firstly with the variety of products that can damage concrete (formed by interaction between cement paste and calcium, sodium, and magnesium sulfates) (Müllauer et al., 2013). Moreover, in the case of sodium sulfate, one school of thought indicates that the ESA mechanism starts with the formation of gypsum, before the one of *expansive* ettringite (Duval & Hornain, 1992). The other school of thought considers the opposite process; once sodium sulfate ions enter concrete, ettringite appears, before the formation of gypsum (Irassar et al., 2003). More precisely, the ESA mechanism, as well as the products formed, seem to change with the solution concentration. For example, according to *Biczok's* results (Biczok, 1967), at low sulfate concentrations, the primary product deposited would be ettringite (before gypsum), while at high concentrations, gypsum would be the main product formed. Hence, for example the ASTM C1012-18 test procedure (2018) or the Canadian one CSA A3004-08 (2008), where the sodium sulfate concentration is 50 g/L (i.e., 5% Na_2SO_4) (see

Sect. 4.4.2), seem from this point of view highly questionable: the favoured mechanisms and thus the degradation potential are different in this case from those prevailing in field conditions.

Likewise, the new performance-based approaches of durability recently developed (see, e.g., Baroghel-Bouny et al., 2004; RILEM, 2007; Alexander et al., 2008) and the recent developments in standardization have induced the need for durability indicators specific to ESA, and/or performance tests carried out directly on concrete samples, in order to directly assess the resistance of concrete mixtures to ESA. Hence, new accelerated tests methods in lab have been developed, in order for example to be more "representative" of field conditions or to suit new binder types. Among the changes applied to the protocol, one can mention the use of **wetting-drying cycles** on concrete samples (Gruyaert et al., 2012; Loser & Leemann, 2016; SIA 262/1, 2019), instead of full immersion of mortar specimens in a sulfate solution.

Hence, the need of improved knowledge has recently pushed people to reactivate the work on this topic. Many PhD theses were defended on this topic (see, e.g., Messad, 2009; Chabrelie, 2010; El Hachem, 2010; Idiart, 2009; Massaad, 2016). In addition, new research projects are still launched in various countries on this topic. The purpose is not only to better understand the chemical and physical mechanisms associated with ESA degradations in field conditions, but also to improve the accelerated lab test methods and define relevant ways to assess the resistance of binders or concretes to ESA.

4.2 Parameters Influencing External Sulfate Attack (ESA) and Accelerated Test Results

4.2.1 Preliminary Remark

Various parameters, such as sulfate concentration of the aggressive solution, counter ion, temperature (e.g., case of thaumasite formation), pH, or degree of saturation, are likely to change the resistance to ESA, or at least the kinetics of reactions, phase change and/or transport. The values of sulfate concentration, counter ion, temperature and degree of saturation are often constant and well known in laboratory testing, whereas they actually vary in field conditions. Conversely, pH is generally constant in outdoor exposure conditions but not systematically controlled in laboratory test procedures. As a consequence it is necessary to know the influence of these parameters on ESA, in order to understand the underlying mechanism(s) associated with the degradation observed in field conditions and to design relevant and "representative" lab test methods.

4.2.2 *Environmental Parameters*

4.2.2.1 Sulfate Concentration: Effect on the Degradation Mechanisms/Products and on the Pore Structure

Sulfate concentration significantly varies in tests procedures, as well as in field exposure conditions. The sulfate concentration of seawater is about 2800 mg/L (Swenson, 1971) and chemically aggressive environments refer to concentrations between 600 and 6000 mg/L (EN 206: 2013), or between 150 and 10,000 mg/L (ACI 318-19, 2019; CSA A3004-08, 2008).

As these conditions should prevail when designing test procedures, the classification based on EN 206: 2013 standard is recalled in Table 4.1.

The sulfate concentration has been proven to be effective in providing significant differences in material performances (see Fig. 4.4). The expansion mechanism is usually assumed to be based on supersaturation: an increase in external sulfate concentration results in an increase in the concentration gradient, and thus faster diffusion and earlier supersaturation of the pore solution. In addition, the maximum concentration of sulfates affects the balance between gypsum and ettringite (Biczok, 1967; Bellmann et al., 2006).

The increase in sulfate concentration up to 30 g/L has been often used in lab testing, in order to accelerate expansion (see, e.g., ASTM C1012-18, 2018). However, a 30 g/L sulfate concentration corresponds to about 10 times the concentrations found in the field. Thus such an increase in sulfate concentration is likely to modify the mechanism of degradation which takes place. As a matter of fact, according to phase equilibrium, the limiting values given for sulfate concentrations from exposure class **XA1** to **XA2** (see Table 4.1) actually correspond to the formation of

Table 4.1 Limiting values associated with exposure classes for sulfate attacks from **natural soil and ground water**, after European standard (EN 206: 2013)

		XA1 Slightly aggressive chemical environment	XA2 Moderately aggressive chemical environment	XA3 Highly aggressive chemical environment
Chemical characteristic	Reference test method			
Ground water				
SO_4^{2-} mg/L	EN 196-2:2013	≥200 and ≤600	>600 and ≤3000	>3000 and ≤6000
Soil				
SO_4^{2-} mg/kg[a] total	EN 196-2:2013[b]	≥2000 and ≤3000	>3000[c] and ≤12,000	>12,000 and ≤24,000

[a]Clay soils with a permeability below 10^{-5} m/s may be moved into a lower class
[b]The test method prescribes the extraction of SO_4^{2-} by hydrochloric acid; alternatively, water extraction may be used, if experience is available in the place of use of the concrete
[c]The 3000 mg/kg limit shall be reduced to 2000 mg/kg, where there is a risk of accumulation of sulfate ions in the concrete due to drying and wetting cycles or capillary suction

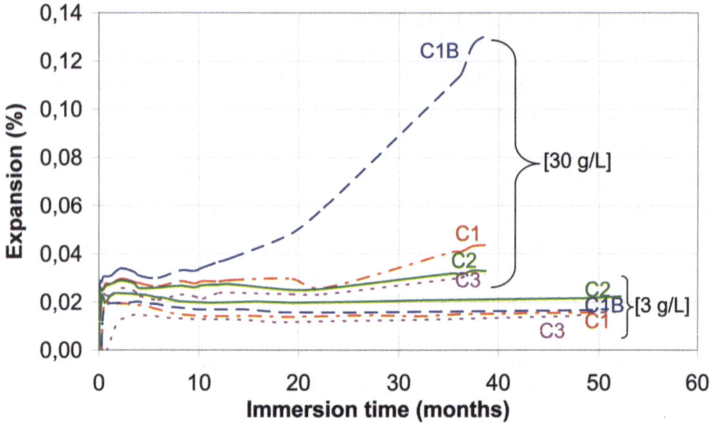

Fig. 4.4 Expansion of 7 × 7 × 28 cm^3 concrete specimens (Rozière et al., 2009). *C3 (W/b = 0.50) contains 30%FA, while C1 and C1B (W/b = 0.70) and C2 (W/b = 0.50) are OPCs*

ettringite. Besides, higher concentrations correspond to the stability of gypsum. As a consequence, the highest sulfate concentrations used in accelerated lab test procedures (30 g/L and above) are likely to result in a degradation mechanism based on gypsum formation, and the degradations then observed are linked to the formation of this product.

However, gypsum can also be found at lower sulfate concentrations (El Hachem et al., 2012b) in the leached outer layer when the pH is not high enough to ensure the Aft stability. Likewise, ettringite can be found in cement-based materials exposed to highly concentrated sulfate solutions.

According to thermodynamic calculation, thaumasite can be formed (in given conditions) from C–S–H and portlandite at very low sulfate concentrations. In the absence of portlandite, higher sulfate ion concentrations are required for the formation of this product (Bellmann & Stark, 2007).

Several studies have been carried out for the last years, in order to study the influence of sulfate concentration on the degradation mechanism. Three concentrations (3, 10, and 30 g/L) have been studied in (El Hachem et al., 2012b; Yu et al., 2012; Müllauer et al., 2013). The accelerating effect was observed on expansion and relative mass curves of mortar specimens (see Figs. 4.5 and 4.6). The three concentrations resulted in the same stages and magnitudes of the phenomenon. Two main stages are usually identified (indicated by arrows on the 3 g/L curve in Figs. 4.5 and 4.6):

- *a first stage* corresponding to the latency (or dormant) period, where no significant expansion can be associated with the relative mass loss caused by leaching,
- *a second step* corresponding to a mass and length increase due to expansion of damaged specimens.

These results are quite similar to those obtained in other studies where these two phases were also identified in the case of mortars specimens immersed in sodium

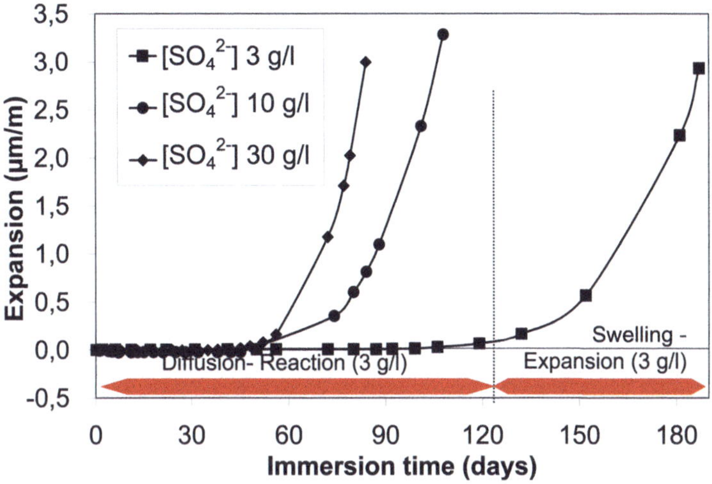

Fig. 4.5 Expansion measured on mortar specimens for 3, 10, and 30 g/L of Na_2SO_4 solutions (El Hachem, 2010)

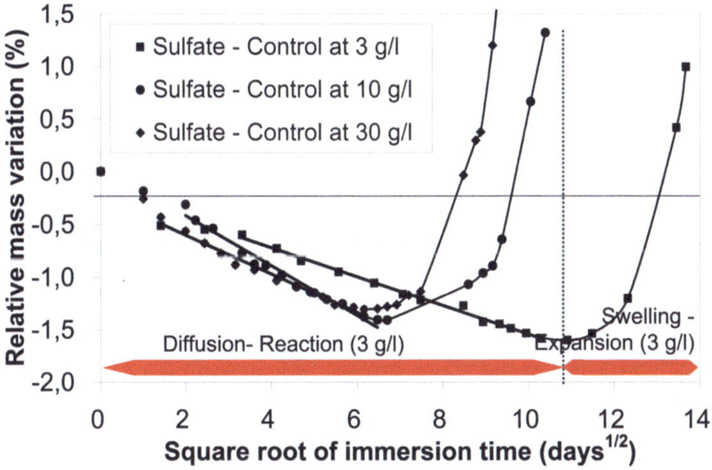

Fig. 4.6 Difference in mass variations between attacked specimens and control (El Hachem, 2010)

sulfate solutions at different concentrations (Lee et al., 2008a; Santhanam et al., 2002a, 2003). It should be noted that these measurements were made on saturated specimens. The mass increase observed can thus be explained by the swelling of the specimens. Taking into consideration the correlation between the mass and length variations, it can be concluded that during the first stage, leaching was the dominant phenomenon and explains the relative mass decrease observed. During the second stage, the formation of expansive products induced expansion and mass increase.

Figure 4.7 shows length measurements as a function of leached ions. At early exposure time, an important quantity of calcium and hydroxide was leached but no expansion was measured. Afterwards, expansion increased with a moderate rate of leached ions. The overall process thus induces calcium ion depletion. Figure 4.7 shows also the common threshold required to trigger expansion. Whatever the concentration, the expansion began at the same level of leached ions. Since the mixtures were the same for the three tests, the calcium hydroxide content was the same at the beginning of the tests. Therefore, the time of initiation of expansion would correspond to the same amount of leached material and ingress of sulfate ions, which could be linked to a critical proportion of the cross section of the specimens required to generate global expansion. If the same critical section results in the same expansion, this suggests that the pressures generated at the different concentrations are the same. This assumption was partly confirmed by the results showed in Fig. 4.8: when the sulfate concentration increased 20 times, the maximum crystallization pressure increased from 7.27 to 8.35 MPa (Müllauer et al., 2013). It was also confirmed in (Bui, 2016) and (Meulenyzer et al., 2020) (see Sect. 4.6.4).

In order to understand the acceleration of expansion observed when the sulfate concentration increases, the microstructure of mortar specimens exposed to three concentrations was analyzed using Scanning Electron Microscope (SEM) mapping coupled with EDS (see Figs. 4.9 and 4.10). The three profiles did not show any significant difference in the sulfate penetration depth. The maximum sulfate content increased with an increase in external sulfate concentration, but the increase was lower than expected, and mainly due to gypsum formation in the cracks. However, these results have several limitations. They only give the sulfate content of the solid phase and not the sulfate concentration in the pore solution. Moreover they give the total sulfate content, but the binding mechanisms (precipitation or adsorption) cannot be distinguished. Hence, the effect of the external sulfate concentration on these parameters is not known.

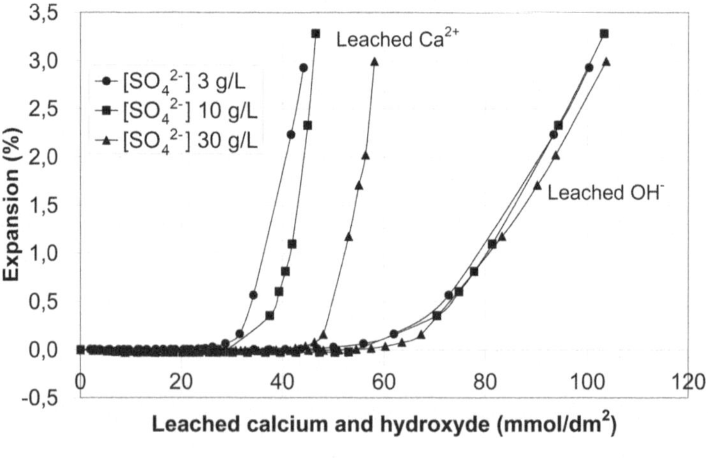

Fig. 4.7 Strain vs. leached ions (calcium and hydroxide) (El Hachem et al., 2012a, b)

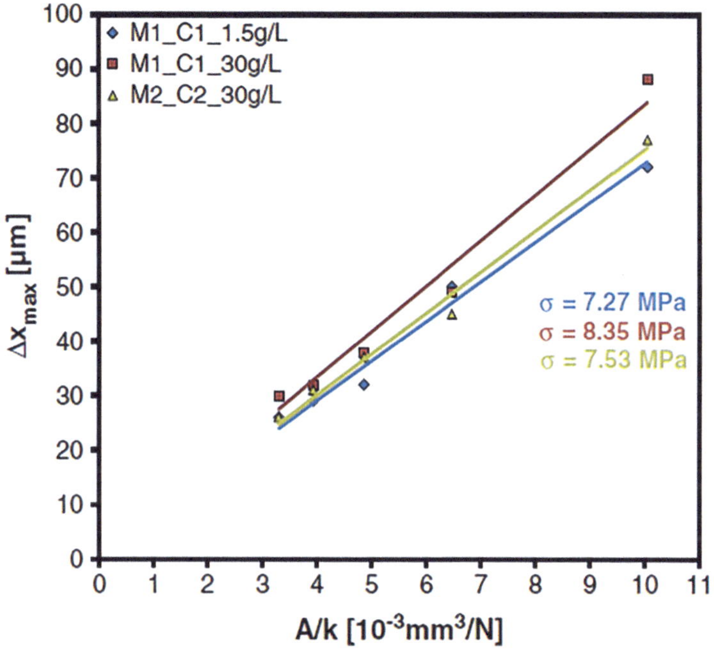

Fig. 4.8 Calculation of maximum crystallization pressure for different cements and sulfate solution concentrations (Müllauer et al., 2013). *M1_C1 and M2_C2 are mortar specimens with W/C = 0.50 and different cements (C1 and C2)*

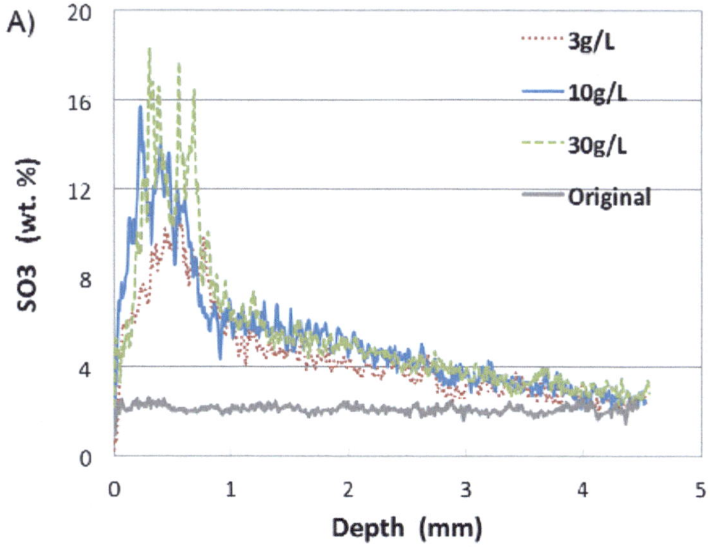

Fig. 4.9 Sulfate content profiles acquired by SEM-EDS mapping, after 120-day exposure to various Na_2SO_4 solutions (Yu et al., 2012, 2013)

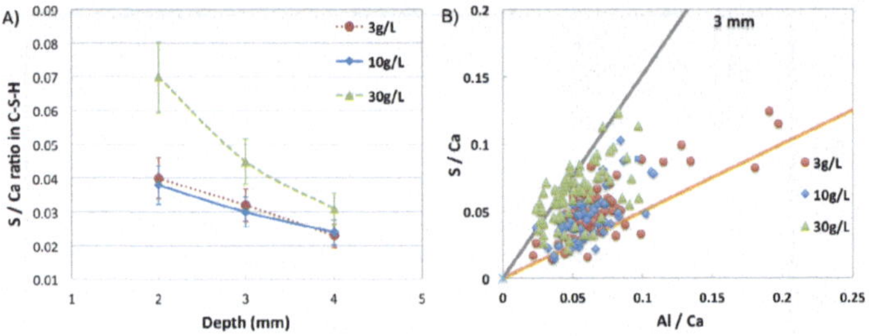

a) S/Ca atomic ratio of the C-S-H vs. depth b) SEM-EDS plots of S/Ca vs. Al/Ca atomic ratios

Fig. 4.10 S/Ca atomic ratio in C-S-H at different depth and SEM-EDS plots of S/Ca versus Al/Ca atomic ratios from the C-S-H at 3-mm depth, for PC mortars in 3, 10 and 30 g/L solutions (Yu et al., 2012). (**a**) S/Ca atomic ratio of the C-S-H vs. depth (**b**) SEM-EDS plots of S/Ca vs. Al/Ca atomic ratios

According to the literature, two mechanisms may explain Aft formation:

- chemical reaction, which involves C_3A,
- surface complexation linked to the capacity of the C-S-H to bind sulfates, which involves $SiO\text{-}CaSO_4$ and depends on the pH value.

Barbarulo et al. (2007) assessed the quantity of sulfate bound to the C-S-H as a function of sulfate concentration of the pore solution. Higher sulfate concentration resulted in higher sulfate content of C-S-H (binding through surface complexation) and ettringite precipitates when a critical sulfate concentration in the solution is reached, which depends on the temperature.

Since surface complexation is assumed to play a major role in the degradation process, quantitative SEM mapping and EDS was also used to assess the sulfate content of the C-S-H (Yu et al., 2012, 2013) (see Fig. 4.10a). This content was higher in the sample exposed to the 30-g/L sulfate solution. The SEM-EDS plots at 3-mm depth (see Fig. 4.10b) actually showed a higher sulfate content of C-S-H in the 30 g/L sample, which corresponds to the formation of ettringite crystals within the C-S-H network. In the 3-g/L and 10-g/L samples, the sulfate content corresponds to monosulfoaluminate crystals mixed with C-S-H. These results help to understand the accelerating effect of high sulfate concentration. At 3-g/L and 10-g/L, the increase in sulfate concentration results in the precipitation of ettringite in the capillary pores, which buffers the increase in sulfate concentration in the pore solution and do not induce any degradation. At 30 g/L, the sulfate content is high enough to transform all the monosulfoaluminate in capillary pores into ettringite, when the sulfate concentration keeps on increasing, this results in ettringite formation in C-S-H pores, and hence swelling and degradation, since according to the *crystallization pressure theory* (Scherer, 2004; Flatt & Scherer, 2008), the finer the pores, the higher the crystallization pressure. As a consequence, ettringite formation (at 3-mm depth) in the smallest pores would explain the higher expansion of the samples exposed to the 30-g/L sulfate solution. Therefore, a higher sulfate concentration favours surface complexation and thus ettringite formation in C-S-H pores. In this case, the growth is controlled by the interface kinetics and not by diffusion. The consequences on the pore structure of the described mechanism has been pointed out thanks to MIP investigations on hardened cement pastes (Gu, 2018; Gu et al., 2019).

4.2.2.2 Counter Ions: Case of Seawater

Sodium sulfate solutions are widely used in performance tests. The solubility of sodium sulfate in water is high (58 g/L). This allows one to make concentrated solutions, in order to benefit from the accelerating effect previously described. Sodium ions do not react with hydration products but their ingress modifies the phase equilibrium of the cement paste and results in an increase in the pH of the pore solution (Guillon, 2004; Guillon & Moranville, 2004).

The solubility of **calcium sulfate** (2.1 g/L) is much lower than sodium sulfate (Duval & Hornain, 1992). The value is used as a reference to design representative tests at $[SO_4^{2-}] = 1.5$ g/L (Planel et al., 2006; Müllauer et al., 2013). The mechanism of degradation is the same; it involves leaching and ettringite formation in the cementitious matrix, which may yield expansion of the test specimens, depending on the experimental conditions.

Magnesium sulfate results in more severe attacks of cementitious materials. They result in spalling, softening, loss of cohesion and significant strength decrease of the test specimens. The degradation is faster than in sodium or calcium sulfate solutions (Duval & Hornain, 1992; Neville, 2004). Magnesium sulfate actually affects all the main hydration products. The dissolution of portlandite produces gypsum and magnesium hydroxide ($Mg(OH)_2$—brucite). The C-S-H are transformed into magnesium silicate hydrates (M-S-H), which have no cohesion. Ettringite has also been observed. It is noteworthy that degradation can occur even at low C_3A and C_4AF concentrations. Such a type of ESA may be encountered in marine environments.

Chloride ions are present in large amounts in **seawater**. The ingress of chloride ions is likely to influence sulfate attack, since it induces the dissolution of monosulfoaluminate and the formation of *Friedel's* salts (e.g., $C_3A.CaCl_2.10H_2O$—hydrocalumite) (Yu & Kirkpatrick, 2001). The released sulfate ions react with available monosulfoaluminate to form ettringite. As a consequence, chloride ingress results in combined ettringite and chloroaluminate formation (Guillon, 2004; Guillon & Moranville, 2004).

In e.g., (Al-Amoudi et al., 1995) and (Lee et al., 2008b), the damages caused by sulfate attack have been found to be delayed and reduced in the presence of chloride ions. Likewise Santhanam et al. (2006) have compared **seawater** exposure to groundwater exposure. Mortar samples were immersed in lime water for 12 days then stored in seawater solutions or sulfate-rich groundwater solutions. Portland cement mortar samples stored in groundwater were more damaged than samples stored in seawater. The weight loss was approximately the same but strength loss and expansion were lower in the case of seawater. Gypsum and ettringite contents were lower in samples exposed to seawater. The authors concluded that chloroaluminate formation and brucite layer precipitation protected the mortar samples against sulfate attack. Such a mitigation of ESA in the case of seawater was taken into account in the French standard (NF P15-317, 2023), where the aluminate content limit was higher than that specified in the case of sulfate-rich environments (see Table 4.1 and EN 206: 2013).

However, combined effects, which take place in structures, should be taken into account. For example, chloride binding, and thus the concentration in chlorides of the pore solution likely to reach the first layer of rebars in RC structures, seems to be affected by the sulfate concentration (Saillio et al., 2016). In addition, the damage, which results from ESA (e.g., increase in porosity and/or microcracking), is likely to accelerate chloride ion ingress. According to Guillon and Moranville (2004) the capillary porosity of cement pastes exposed to sodium chloride first slightly decreases due to ettringite formation then increases due to portlandite dissolution. Both effects will contribute to enhance the risk of reinforcement corrosion.

Note that a few studies deal with the combined effects of sulfate and chloride ions (see, e.g., Xu, 1997; Guillon, 2004; Zuquan et al., 2007; Maes & De Belie, 2014; Saillio et al., 2016). For example, *Saillio* et al. observed that chloride ingress is enhanced in concrete lab samples when both chloride and sulfates are present in the contact solution at pH = 8, whatever the binder type (Saillio et al., 2016). On the other hand, concrete stored in chloride and sulfate solutions for 12 months showed lower chloride diffusivity in (Feldman et al., 1991). According to De Weerdt et al. (2019), phase changes due to the presence of sulfur and magnesium in seawater did not limit chloride ingress compared to the case of NaCl exposure, in the conditions/time of exposure applied. Experimental studies reported in (Zuquan et al., 2007) showed that chloride ingress is first reduced in sulfate solution (3.5% NaCl and 5% Na_2SO_4) but is accelerated during the second stage. On the basis of (Dehwah, 2007), initiation of steel corrosion in concrete exposed to chloride ions is not influenced by the addition of sulfate ions, but the corrosion rate is found to be dependent on the sulfate concentration.

The role of alkalis still needs also to be clarified.

Hence, even if experimental lab studies have provided data on external sulfate attacks combined with chloride ingress, it is very difficult to predict the behaviour of concrete structures under the combined action of sulfates and chlorides in field conditions. The results strongly depend on the composition of both the material and contact solution, and on the time of exposure.

The solubility of each component of the sulfate solution is influenced by the others. For example, the solubility of calcium sulfate is higher in a sodium chloride solution. Therefore, it is necessary to take into account the combined effects of ions in order to predict the performance of cementitious materials in field exposures.

4.2.2.3 Temperature of the Environment

The effects of temperature on external sulfate attacks are not fully understood. However this issue can be divided into two parts: low temperatures and high temperatures.

The influence of low temperature has been widely investigated, since it has been assumed to favor **thaumasite formation** (Barker & Hobbs, 1999; Hobbs & Taylor, 2000; Crammond, 2002; Blanco-Varela et al., 2006; Pipilikaki et al., 2008; Schmidt et al., 2008). Thaumasite has been found in samples immersed in a sodium sulfate solution at T = 5 °C. At T = 10 °C thaumasite and calcite were found, whereas at T = 20 °C, thaumasite could not be detected. In addition, it has been reported that thaumasite could also form at ambient temperatures, depending on the conditions (Irassar et al., 2003). This is quite possible according to thermodynamical considerations. Thaumasite formation results in lower expansion but in decrease in strength at T = 5 °C (see, e.g., Figure 4.11). Thaumasite can form from C-S-H and sulfates in the presence of calcite (Schmidt et al., 2009).

Understanding the influence of high temperature is also a way to understand the mechanisms of Delayed Ettringite Formation (DEF), which causes significant damages in steam-cured concrete (heat treatment of prefabrication plants) or in large concrete structures where the temperature exceeds T = 80 °C at early age due to heat released by hydration (Scrivener & Taylor, 1993; Fu et al., 1994; Lawrence, 1995; Divet & Randriambololona, 1998; Lothenbach et al., 2007). High temperature is also used in accelerated lab tests (Aköz et al., 1999), in particular to be representative of exposure conditions of immobilization systems (Hernández et al., 1997; Santhanam et al., 2002b). Barbarulo et al. (2007) have studied the chemical equilibrium between C-S-H and ettringite, at T = 20 °C and T = 85 °C. The quantity of sulfate bound to the C-S-H is not negligible in a cement paste. It increases with sulfate concentration and temperature (Divet & Randriambololona, 1998; Barbarulo et al., 2007). The solubility of ettringite increases with temperature, from around 0.4 mmol/L of sulfate in solution at T = 20 °C to 5 mmol/L at T = 85 °C. At high temperature (during hydration or steam curing), ettringite becomes more soluble, the sulfate concentration in solution increases and the quantity of sulfate bound to the C-S-H increases accordingly. When the system is placed at room temperature, ettringite can precipitate, but the kinetics is quite slow. Increasing temperature resulted in a decrease in the latency period, but the expansion rate and the magnitude were not modified (see Fig. 4.12 (Santhanam et al., 2002b)).

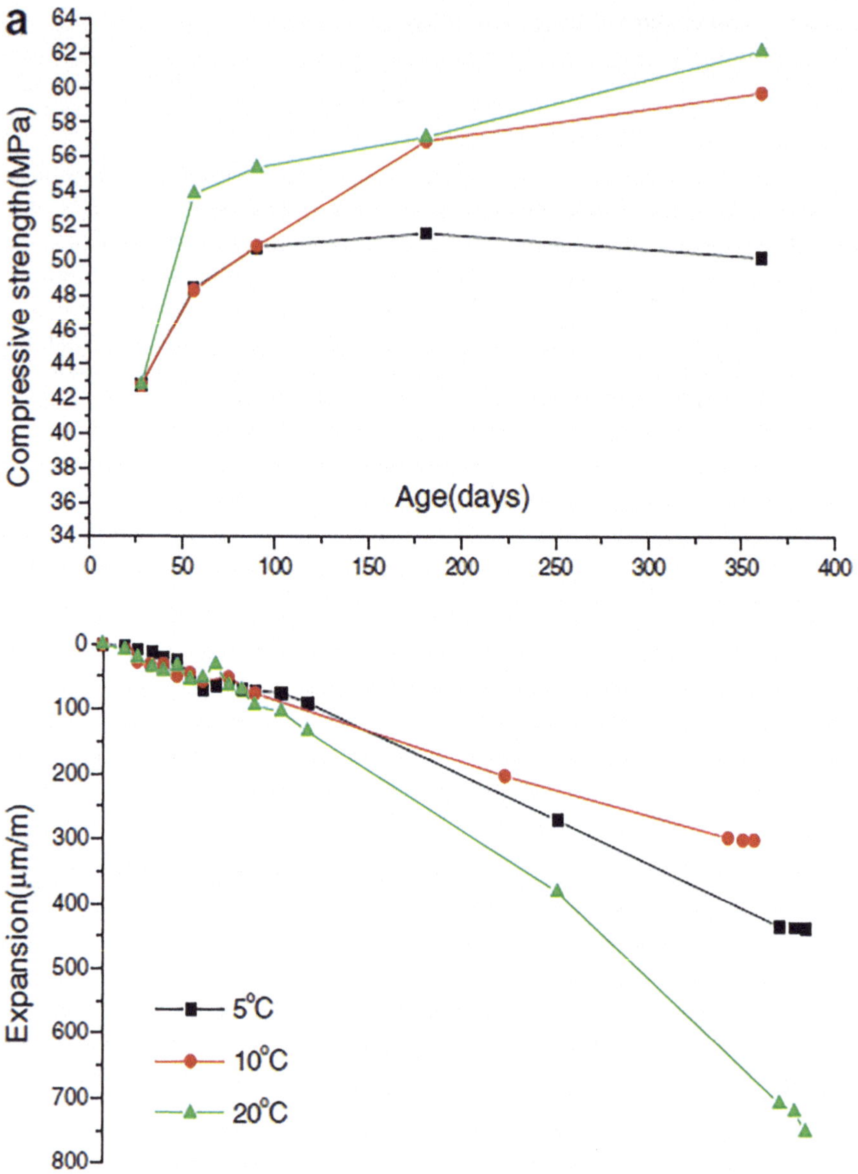

Fig. 4.11 Effect of the temperature of storage in a 5%-Na_2SO_4 solution on compressive strength and expansion (Pipilikaki et al., 2008)

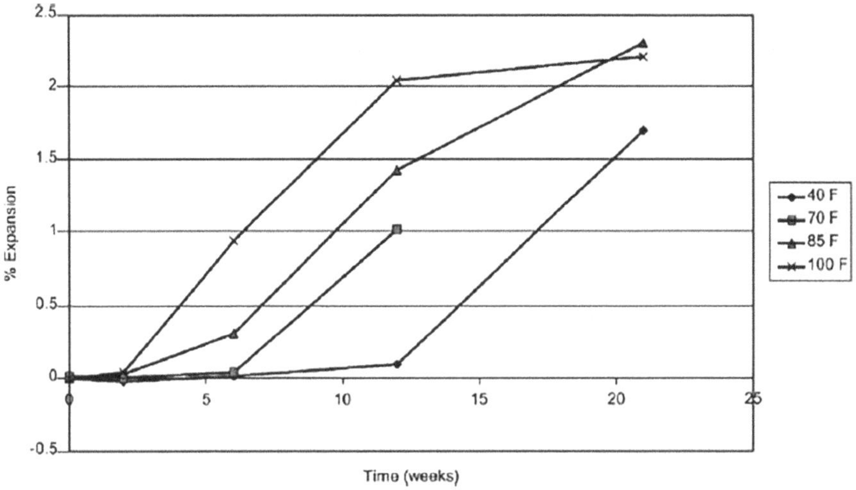

Fig. 4.12 Expansion of Portland cement mortars immersed in sodium sulfate solutions at various temperatures (Santhanam et al., 2002b)

4.2.2.4 pH of the Contact Solution

A continuous supply of sulfate solution and constant pH can be assumed in field conditions. More precisely, constant pH close to 8 can be assumed in the case of seawater exposure conditions.

The pH can be kept to a value close to 8 for ESA lab tests, in order to be more "representative" of seawater exposure conditions.

Most of experimental lab procedures recommend regular renewal of sulfate solution, in order to keep sulfate concentration constant, but this does not ensure constant pH. As a matter of fact, as a result of leaching of alkalis and calcium, the pH increases up to 12 or 13 within a few hours after contact between the material and the solution. The pH can be kept constant by adding concentrated acid solution (Mehta, 1973, 1975). An example of setup is presented in Fig. 4.13. One pH controller can be used for several concrete or mortar mixtures, and titration of the solution is not required. Concrete or mortar samples are immersed in sodium sulfate solution. The sulfate concentration is 3 g/L, and pH is controlled using 0.02-mol/L sulfuric acid solution (Rozière et al., 2009). Solutions are renewed every month. Mass, length and modulus elasticity of concrete samples are monitored.

Keeping the pH constant accelerates the test (Brown, 1981), as shown by the loss of strength and the expansion rate. For example, comparative tests have shown that the time needed to reach significant expansion for the uncontrolled pH condition is roughly twice as much as for the controlled pH condition (Brown, 1981; Ferraris et al., 1997). Hence, pH control can result in shorter tests.

But in this case, coupling between "pure" ESA and leaching of alkalis and calcium takes place. This increases the difficulty in the analysis and interpretation of the test results.

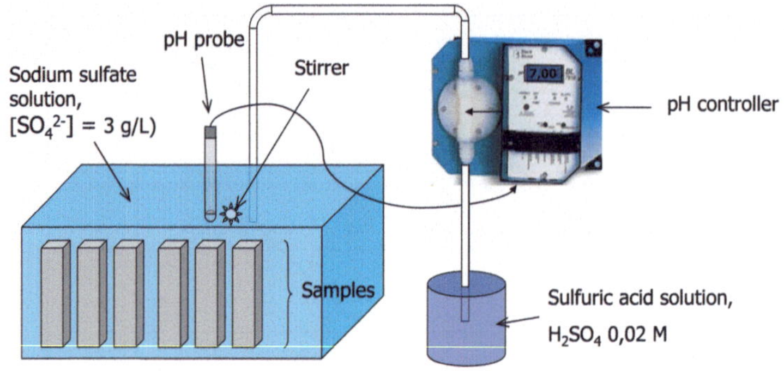

Fig. 4.13 Setup of accelerated lab test with pH control (Rozière, 2007)

4.2.3 Degree of Saturation of the Material: Case of Drying-Wetting Cycles

4.2.3.1 Field Experience

Wetting and drying cycles may affect concrete structures, not only in tidal or splash zones in marine environments, but also in foundations and buried structures, since the level of water in soils, rivers, and sea is likely to vary. Consequently, the degree of (liquid water) saturation of concrete exposed to sulfates in these environments will also vary.

Drying of concrete may yield crystallization of salts (see Sect. 4.1.1). Damages caused by the precipitation or the crystallization of sodium sulfate salts and referred to as "physical sulfate attacks". They actually do not result from chemical reactions between sulfate and cementitious materials. Sodium sulfate exists as an anhydrate called thenardite (Na_2SO_4), or as a decahydrate called mirabilite ($Na_2SO_4.10H_2O$). Wetting (immersion) and drying cycles cause repeated thenardite dissolution and recrystallization and volume changes and thus fatigue in specimens. Stresses sufficiently high to overcome the tensile strength of the material may be developed and the salt may exert a sufficient force to bring about disintegration of the material (Bonnell & Nottage, 1939). Accelerated degradation of the hydration products, as a result of wetting and drying cycles has been reported in the literature (see, e.g., Bassuoni & Nehdi, 2009b; Gao et al., 2013).

Hence, the occurrence of drying-wetting cycles is likely to completely change the underlying ESA mechanisms.

4.2.3.2 Lab Tests

Ingress of aggressive species by capillary absorption is known to be faster than diffusion (François et al., 2001). Hence, initial drying (e.g., preliminary drying at T = 60 °C) and re-saturation, by the same sulfate solution as used in the immersion test, has been used to accelerate ESA in some lab test protocols (see, e.g., Messad, 2009; El Hachem, 2010).

However, most of the experimental procedures involve full immersion of concrete or mortar specimens in sulfate solution. As a matter of fact, this is representative of a great number of exposure conditions and provides more robust procedures than wetting-drying cycles (Chabrelie, 2010).

The effect of drying on the initiation time of expansion has been studied on OPC-0.60 mortar (El Hachem, 2010). Mortar prisms were stored in lime-saturated solution for 6 months. Then two samples were directly immersed in sodium sulfate solution ($[SO_4^{2-}] = 3$ g/L), two other samples were stored at 20 °C and 50% RH, and two samples were oven dried at 45 °C until mass stabilization. The final bulk water saturation degrees of the samples '6 months', '6 months + 20 °C' '6 months + 45 °C' were respectively 100%, 67% et 1%. After drying at 20 °C and 45 °C, the mortar samples were immersed in sodium sulfate solution. The pH of the sulfate solutions was controlled and the amounts of added acidic solution were monitored. Mass and length of samples were measured each time the sulfate solution was renewed (see Figs. 4.14 and 4.15).

Figure 4.14 shows the mass evolution of the specimens during the drying period in the left part of the graph. The reference time (t = 0) corresponds to immersion. After immersion in sodium sulfate solution, there was a noticeable mass increase in the dried specimens. This is due to re-wetting of the specimens and not to sulfate attack. It was actually equal to the mass lost while drying.

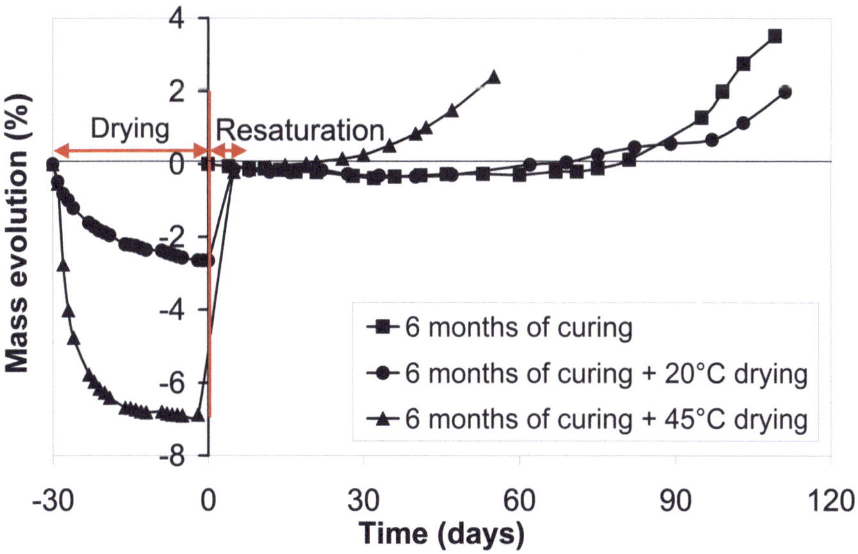

Fig. 4.14 Mass change vs. time for three different curing procedures (El Hachem, 2010)

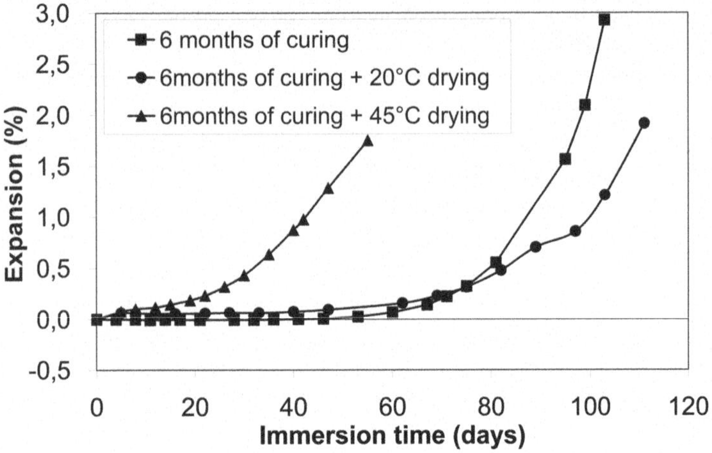

Fig. 4.15 Expansion vs. time for three different curing procedures (El Hachem, 2010)

The shape of the expansion curve presented in Fig. 4.15 is quite similar to that of mass increase (except for the first day) (see Fig. 4.14). In addition, expansion was observed at the same time as mass increase.

A latency period was clearly observed for the specimens directly immersed in the sulfate solution and those dried at 20 °C. It was almost not noticeable for those dried at 45 °C. The mass increase due to sulfate attack started very early (<20 days) in the specimens dried at 45 °C. Upon immersion, the sulfate solution penetrates (after drying) mainly by capillary absorption. This explains the rapid response. The specimens directly immersed in the sulfate solution and those dried for one more month at 20 °C had the same response approximately during the first 2 months before mass increase. The sulfate ions penetrate mainly by diffusion in this case. At the beginning of mass increase, the first samples react faster than the second ones and the degradation is more obvious. The difference between these two behaviors could be related to the pore size distribution involved and the degree of saturation of the sample (see Fig. 4.16). The ingress of sulfate ions by capillary absorption is easier/faster into the pore system after drying at 45 °C than after drying at 20 °C. In addition, the poor resistance of the samples dried at 45 °C could be attributed to the consecutive wetting and drying cycles that can cause fatigue and generate microcracking of the cementitious matrices. As a matter of fact, this leads to more permeable systems (Bassuoni & Nehdi, 2009a, b). Preliminary drying hence results in faster ingress of sulfate ions during immersion and significant acceleration of sulfate attack. This principle was used by *Messad* et al. (Messad, 2009; Messad et al., 2010; Linger et al., 2014), in order to design accelerated ESA lab tests on concrete samples (see Sect. 4.5.1). However, attention should be paid to the influence of the temperature increase on the microstructure of cementitious matrix and on the possible occurrence of DEF-based mechanisms (see Fig. 4.17) (Lothenbach et al., 2007; Lothenbach et al., 2010).

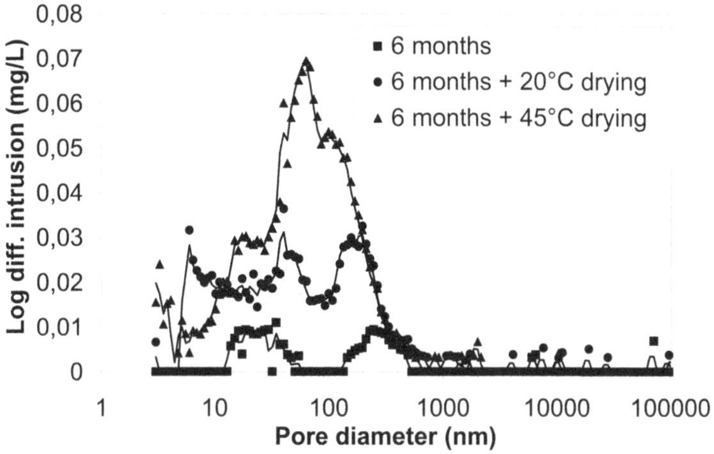

Fig. 4.16 Influence of initial drying on pore size distribution. (After El Hachem, 2010)

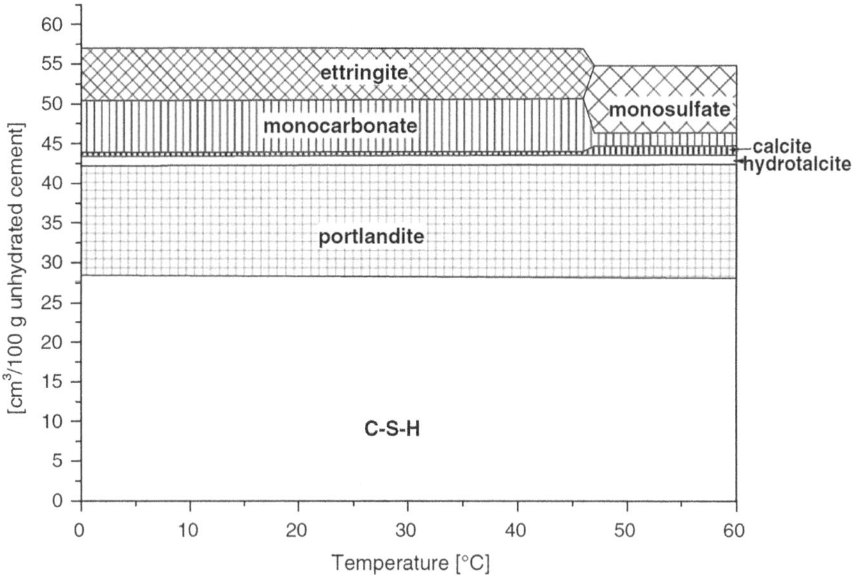

Fig. 4.17 Calculated volume of solids in a hydrated cement paste as a function of temperature. (Adapted from Lothenbach et al., 2007, 2010)

4.2.4 Mix-Parameters: Case of Low-CO_2 Binders

The influence of the binder type has been widely studied. In particular, in the last decades, a focus on the effect of incorporation of SCMs such as FA or GGBS on ESA has been made (see, e.g., Tikalsky & Carrasquillo, 1989; Sahmaran et al., 2007; Chabrelie, 2010; Gruyaert et al., 2012; Ramezanianpour & Hooton, 2013; Yu et al., 2013; Hossack & Thomas, 2015; RILEM, 2018; Jabbour et al., 2019). Not only these (mature) low-CO_2 binders are usually revealed more durable than OPC-concretes for some other aspects of durability such as chloride ingress or alkali-reaction, but also an enhanced performance was reported in some recent works for alkali-activated ground granulated blast furnace slag (GGBS) and binary blends against ESA. For example, CEM-III mixtures with high volume of GGBS (e.g., 60%) usually reveal better behaviour than OPC-mixtures with regard to resistance to ESA in full immersion and are usually considered as sulfate-resistant in this case. In addition, mixtures with 30% or 50% GGBS were found sulfate-resistant at 23 °C, on the basis of the CSA A3004-08 Canadian test procedure (2008) (similar to ASTM C1012-18, 2018) (Ramezanianpour & Hooton, 2013). This may result from formation in GGBS mixtures of a high quantity of **hydrotalcite**, which has a high capacity to bind the sulfate and aluminate ions released during ESA (Schmidt et al., 2008, 2009; Khan et al., 2017).

In the case of FA, short (e.g., 3-day) moist curing before exposure to sulfate-rich environments yield ESA-induced deterioration, since little of the FA has reacted after such a short curing.

Moreover, it is worth reminding that the presence of **limestone** in the mixture enhances thaumasite-induced degradation at low temperature, as observed both in lab and field conditions (see, e.g., Ramezanianpour & Hooton, 2013; Hossack et al., 2020).

4.2.5 Test Specimens

In saturated conditions, the mechanism of degradation involves sulfate ion ingress in the cementitious matrix by diffusion (possibly coupled with leaching), thus the characteristic time of immersion tests and in particular the time required to observe damage, as well as the degradation mode of the samples, depends on the diffusivity and thus on the size (e.g., volume of the specimen and exposed surface) and geometry of the specimens.

The effect of the shape/dimensions of the specimens has been studied in the literature (see, e.g., CEN/TR 15697, 2008; El Hachem, 2010; Yu et al., 2013; Massaad et al., 2016b; Jabbour et al., 2019; Suma & Santhanam, 2020).

Solid prismatic and cylindrical specimens are the most used in the tests. In order to speed up and enlarge the response of the samples exposed to ESA, hollow cylindrical specimens can be used (Massaad, 2016; Massaad et al., 2016a, b).

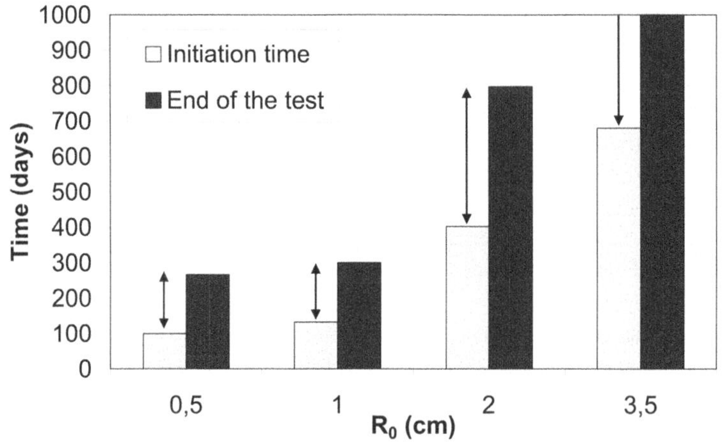

Fig. 4.18 Times of initiation (for expansion) and of end of the test. Mortar specimens immersed in 3-g/L sodium sulfate solution (El Hachem, 2010)

With the use of small specimens, the relative volume of degraded material, and/or which swells, reaches earlier the critical detection threshold and the values measured are larger. This last aspect may improve the precision of the measurement. However, edge effects can also be enhanced.

The drying radius $R_0 = 2$ V/S (V is the specimen volume and S is the exposed surface) can be used to represent the influence of the specimen size. Shrinkage half-time is a linear function of R_0^2, according to EN 1992-2: 2005. Smaller specimens, i.e., with relatively small drying radius, are the most reactive specimens to ESA (El Hachem et al., 2012b). More precisely, the larger the specimen, the longer the latency period (or initiation time) when significant expansion can be measured. The influence of the drying radius on initiation time is less strong (see Fig. 4.18), which suggests that the wall effect plays a significant part. The skin of mortar specimens delays the consequences of ESA in particular on the smallest specimens.

Suma and Santhanam (2020) have investigated the role of mortar specimen size on the rate of length change in sodium sulfate attack and indicated that smaller sizes than ASTM C1012-18 standard size of 25 × 25 × 285 mm^3 (2018) could be used, in order to accelerate the test. As a matter of fact, in this case the sulfate penetration depth will be larger and thus will affect a larger portion of the specimen while the restraining effect of the non-affected core will be reduced.

In order to study the effect of specimen configuration, an experimental campaign has been carried out and is reported in (Massaad et al., 2017). 8 tested specimen configurations, 5 geometries (prismatic, cylindrical and 3 hollow cylindrical geometries) and 3 compositions (concrete, concrete-equivalent mortar and standard mortar) were tested. The cylindrical mortar specimen was selected by the authors as the most adapted, in term of ESA degradation expression, for the adopted performance-based approach. The different configurations lead to the same overall degradation mechanism, but the geometry and the composition affected the magnitude of some chemical or physical parameters, due to couplings between chemical and mechanical processes (Massaad et al., 2017).

Representative volume of **mortar** requires small specimens. For example, 4-mm maximum aggregate size results in 20-mm thick specimen and seven times lower initiation time than 70-mm thick specimen (required for concrete specimens). The ASTM C1012-18 (2018) standard prismatic mortar bars have a cross section of 25×25 mm^2. This allows investigating the effect of several mix-design parameters, such as binder type and water-to-binder ratio, and designing performance-based specifications. However shifting from mortar to concrete may have consequences on resistance to ESA that are still not completely understood.

Tests on **concrete** specimens (instead of mortar) are more appropriate in view of optimizing concrete mix-design or evaluating the performance/durability of concrete structures. However representative volume of concrete results in relatively large specimens. A thickness of more than five times the maximum aggregates size is commonly recommended. This results in relatively long initiation times (see, e.g., Figure 4.18), exceeding several years. This has been shown for example trough the long-term investigation of the behavior of concrete exposed to sodium sulfate conducted by the US Bureau of Reclamation in the 1950s. In this case, the specimens were continuously exposed to sulfates over a period of more than 40 years (Neville, 2004).

4.3 Usual ESA Accelerated Test Methods: General Features, Issues and Measured Parameters

4.3.1 General Features and Issues

Among the issues discussed in the research community, there is the question of "accelerated" test applied on cementitious material samples in lab conditions. Various experimental works, which were carried out in order to (further) accelerate sulfate attack in lab, are reported in the literature. However, the majority requires a long duration of the test (between 6 months and 1 year). On the other hand, there was a need from the practice for a more rapid test (no longer than 12 weeks).

Owing to the slowly ESA process occurring in the field, the way to understand and analyze the reaction mechanisms, as well as to evaluate resistance to ESA, has consisted for decades in carrying out accelerated tests in lab, by:

- using **high sulfate solution concentrations** (e.g., >30 g/L), in order to enhance the transport process,
- storing at **high temperature** (e.g., >30 °C), in order to promote the reaction kinetics (except in the case of the thaumasite-form of attack),
- **reducing the curing time**, in order to obtain a more porous material and enhance the transport process (Chabrelie, 2010).

The last point (curing time) sounds in adequacy with the real conditions of the structures in the field, since usually concrete structures are exposed to various physical and chemical aggressions as soon as their formworks are removed. In addition, such an early-age exposure has an obvious influence on the transport properties of the material in lab and this will accelerate the ESA process, as previously mentioned. Nevertheless, in this case the performance of SCM-materials may be underestimated.

One the other hand, the two other points are not in adequacy with real conditions: the high concentration usually used in accelerated tests corresponds to about 10 times the maximum values found in the field and this can change the degradation mechanisms (e.g., enhanced formation of "secondary" gypsum). For example, the sodium sulfate concentration in ASTM tests (ASTM C452-15 (2015) and ASTM C1012-18 (2018)) and in Canadian tests (CSA A3004-08 (2008)) is 50 g/L. In the case of seawater, the sulfate concentration is about 2.7 g/L and in ground water or soils, this value can be much lower. Likewise, typical water temperature in the field reaches 10 to 20 °C, which is significantly less than that used in accelerated tests.

Hence, the existing ESA test methods used to evaluate the performance of cementitious materials in sulfate-rich environments and described in the literature are often criticized, not only for their slowness, but also for their exposure conditions (Brown, 1981; CEN/TR 15697, 2008; Cohen & Mather, 1991; Mehta, 1975; Planel et al., 2006; Skalny & Pierce, 1999). This is the case of the pH of the immersion solution. As a matter of fact, its value is changing as a function of time if the solution is not periodically renewed (Mehta, 1975). For example, as a result of leaching, the pH changes from close to neutral to about 12.5 (equilibrium pH of the pore solution) within a few hours. This can be considered as far from field conditions where the reactants are inexhaustible and always accessible. In addition, the aggressiveness of the solution will decrease as a function of time and this will lengthen the test with respect to the case where in particular the sulfate concentrations is constant (Mehta, 1975; Revertegat et al., 1992).

In order to improve the methods and in particular to accelerate further the process, some authors suggested, e.g.:

- adapting the specimen size,
- modifying the sulfate ingress process (capillary absorption vs. diffusion). In saturated conditions, the testing time actually depends on the diffusion characteristic time,
- maintaining the pH of the solution at a constant value (e.g., pH = 8 as in seawater) with a periodic renewal of the sulfate solution (Mehta, 1975; Brown, 1981; Santhanam et al., 2001). As a matter of fact, *Brown* in (Brown, 1981) showed that the expansion rate increases when the pH of the immersion solution is controlled to values lower than 11.5. Nevertheless, the ESA mechanisms are changed also in this case. Thus, when the pH value is between 8 and 11.5, gypsum is formed and C-S-H are decalcified, and this affects the material cohesion (Khatri et al., 1997).

Most of the available (standard) test methods deal with the *ettringite form of ESA*. Note however that the Canadian standard CSA A3004-C8 (2008) is dedicated to the *thaumasite form of sulfate attack* (see Sect. 4.4.2).

4.3.2 Measured Parameters

Numerous tests have been designed to assess the performance of cementitious materials exposed to ESA. The **parameter** usually measured is longitudinal expansion (during the immersion test). In various studies, other concrete properties are measured before and after immersion, such as mechanical properties (compressive strength, according to EN 12390-3: 2019), tensile strength and/or E-modulus (Loser & Leemann, 2016), chloride diffusion coefficient by various methods, and bulk porosity (NF P18-459, 2022). Visual observations, as well as mass loss measurement, are also often carried out, along with 1-D deformations. Resistivity and pH measurements are scarcer.

It is worth mentioning here that, in current standards and recommendations, acceptation of a cement as sulfate resistant is based on expansion measurement.

4.4 Existing Standards or Recommended Test Methods in Various Countries for Sulfate Resistance Testing

Test methods dedicated to ESA have been standardized and are currently applied in various countries. Various widely used test methods for ESA testing are detailed in the following sections. Some of them are also summarized in Table 4.2. In the methods described in the following, most of them involve full immersion, one uses initial contact (during mixing) with sulfates (ASTM C452-15, 2015), and some others involve drying-immersion cycles.

4.4.1 Chinese Standard Test Methods

With regard to the assessment of the resistance to ESA, various standards are used in China:

- **GB/T 749-2008**: Test method for determining capability of resisting sulfate corrosion of cement (2008)
- is used for **cement**. Its main features are the following:
  - $10 \times 10 \times 60\ mm^3$ mortar prisms are cured in water at 50 °C for 7 days,
  - the mixtures are characterized by S/B = 2.5 and W/B = 0.5,
  - the cured samples are immersed in 3% Na_2SO_4 solution for 28 days,

Table 4.2 Comparative summary of some widely used test methods for ESA testing (full immersion)

Test & reference/ Scope	Mix proportions and Storage	Specimen geometry and size (mm)	Measured parameters/Time required/Maximum acceptation range	Limits of the method
ASTM C452-15 (2015) Measurement of potential expansion of Portland cement mortar bars. Dry OPC premixed with gypsum with a SO_3 content of 7.0% by mass	Cast with water:cement:fine aggregate ratio 0.46:1:2.75 for air-entrained mortar with Portland cement and 0.485:1:2.75 for non-air entrained mortar. Curing at 23 ± 2 °C in water at least for 30 min before first measurement. Storage of specimens in water with volume of water/sample not exceeding 5 to prevent excessive leaching	Mortar bars with gauge studs 25 × 25 × 285 mm^3	Length measurement: 14 days; first measurement after 24 hours 0.01% for 3 samples, 0.011% for 4 samples, 0.012% for 5 samples, 0.012% for 6 samples	Not intended for mixtures with SCMs, due to premixed gypsum
ASTM C1012-18 (2018) Measurement of length change of hydraulic cement mortar bars exposed to sodium sulfate solution	Cast with water:cement:fine aggregate ratio 0.46:1:2.75 for air-entrained mortar with Portland cement and 0.485:1:2.75 for non-air entrained mortar. For other cements maintain flow as 110 ± 5 by flow table test. Maintain 35 ± 3 °C for 24 hours in water bath. After demolding, curing of specimens until the specimens reach a compressive strength of 20 ± 1 MPa in lime water, then immersion in 50 g/L sodium sulfate solution. pH range 6 to 8. volume of solution/sample is kept as 4	Mortar bars with gauge studs 25 × 25 × 285 mm^3	Length measurement: 6 months to 1 year. Readings at 1, 2, 3, 4, 8, 13 and 15 weeks, later at 4, 6, 9, 12, 15 and 18 months. For blended cements: 0.034%, 0.037%, 0.039% and 0.041% respectively for 3, 4, 5 and 6 specimens. For Portland cements: 0.01% for 3 samples, 0.011% for 4 samples, 0.012%for 5 samples, 0.012% for 6 samples	At least 6 months are required to conclude the test

(continued)

Table 4.2 (continued)

Test & reference/ Scope	Mix proportions and Storage	Specimen geometry and size (mm)	Measured parameters/Time required/Maximum acceptation range	Limits of the method
(German) **SVA test method** (DIBt, n.d.)	Cast mortar with W/C ratio of 0.5. Specimens demolded after 2 days and subsequently stored for 12 days in saturated lime water Immersion in 4.4% sodium sulfate (Na_2SO_4) solution (29.8 g of sulfate in 1 litre water). Storage: 91 days *Note that the **modified SVA test method**, which is currently used, has a 0.44% sodium sulfate solution and lasts at least 2 years*	Flat mortar bars $10 \times 40 \times 160$ mm^3	Length measurements: ***SVA test method*** - *4.4% sodium sulfate solution*: 14, 28, 56 and 91 d storage time $\Delta L/Lo \leq 0.50$ mm/m after 91 days **Modified SVA test method** - *0.44% sodium sulfate solution*: every 13 weeks. Comparison of the results with CEM I-SR and CEM III/B-SR mortars	The *initial version* leads to very high length changes and often failures of CEM I-SR The *modified version* takes at least 2 years to conclude the test

- the capability of resisting ESA is assessed from the ratio of flexural strength of the corroded mortar sample on the non-corroded one.

- **GB/T 50082-2009**: Standard for test methods of long-term performance and durability of ordinary concrete (2009)
- is used for **concrete**. Its main features are the following:
  - $100 \times 100 \times 100$ mm^3 cubic samples cured for 26 days in standard conditions and then dried at 80 °C for 2 days,
  - the samples are immersed in 5% Na_2SO_4 solution for 15 hours at 20 °C, then dried at 80 °C for 6 hours and cooled down in 2 hours,
  - the compressive strength of concrete is tested after every 25 cycles,
  - the capability of resisting ESA is assessed from the strength ratio of the corroded concrete sample on the uncorroded one.

The sulfate resistance factor of concrete (KS) is then assessed:

$$KS = f_{corr} / f_{uncorr} \left(in\%\right)$$

4.4.2 Test Methods in North America

Three standard tests are used in the USA, in order to evaluate cement or concrete performance with regard to sulfate attack:

- **ASTM C452-15** "Standard Test Method for Potential Expansion of Portland-Cement Mortars Exposed to Sulfate" (2015),
- **ASTM C1012-18** "Standard Test Method for Length Change of Hydraulic-Cement Mortars Exposed to a Sulfate Solution" (2018),
- **USBR 4908-86** "Procedure for Length Change of Hardened Concrete Exposed to Alkali Sulfates" (1986).

Likewise, two standard tests are used in Canada, in order to evaluate cement performance with regard to sulfate attack:

- **CSA A3004-C6** "Test Method for Determination of Mortar Bar Expansion Due to External Sulphate" (2008),
- **CSA A3004-C8** "Test Method for Determination of Sulphate Resistance of Mortar Bars Exposed to Sulphate Solution" (2008).

Numerous research works have used these standard test methods or variants of them, for example in order to study the effect of various parameters or with the purpose of establishing performance-based specifications for the durability of concrete against external sulfate attack (see, e.g., Stephens & Carrasquillo, 2000; Ramezanianpour & Hooton, 2013; Hodhod & Salama, 2014).

The respective test procedures are detailed in the following sections.

4.4.2.1 ASTM C452

The **ASTM C452-15** test method (2015) implies measuring the level of expansion (induced by ettringite formation) in a mortar bar specimen cast with Portland cement and gypsum, which is considered here as the (internal) source of sulfates.

Therefore, this test method does not directly deal with external attack. In particular, this method does not take into consideration the transport properties of the material tested, as well as the resistance to (sulfate) penetration. In addition, the presence of gypsum in the mix may affect the whole hydration process (e.g., hydration kinetics and morphology of the C-S-H formed (Mota et al., 2015)), as well as yield reactions with anhydrous compounds and induce degradation at early stage. This is different from the field conditions of occurrence of ESA.

However, this test method is listed in ASTM C150-18 (2018) as a way of evaluating cement materials attacked by sulfates. According to ASTM C150-18, the maximum level of expansion accepted is 0.04% at 14 days for severe sulfate resistant (type V) cements (and 0.06% for moderate sulfate resistant (type II) cements).

The ASTM C452-15 test method (2015) is known to be fast, since the resistance to ESA can be assessed in this case **after 14-day exposure**. This is its main (single?) advantage of this test method. It is also recognized as efficient to differentiate high-C_3A and low-C_3A OPC cements with regard to ESA susceptibility.

This test can be considered somehow inaccurate when the mortar mixture contains SCMs. This can be explained by the fact that SCMs have not sufficiently reacted within the 14-day (curing) period where the expansion is measured. Consequently, the benefit of SCMs cannot be clearly identified in this test. This test should hence only be applied to test Portland cement.

The test procedure is described in Table 4.2.

4.4.2.2 ASTM C1012

Given the limitations of ASTM C452-15 test method (2015) previously mentioned, the **ASTM C1012-18** test method (2018) was developed, in order to be applicable to mixtures containing SCMs.

In ASTM C1012-18 test method (2018), mortar specimens are immersed in a sulfate solution once they reach a certain level of compressive strength. This required level is around 19.7 MPa. The specimens are $28 \times 28 \times 285$ mm^3 bars. The concentration of the sulfate solution used in this test is 0.352 $mol.L^{-1}$, which is equal to 50 g/L sodium sulfate.

With regard to this test, the maximum level of expansion accepted is 0.05% after 180-day immersion, for severe sulfate resistant (type V) cements (and 0.1% for moderate sulfate resistant (type II) cements).

The results obtained when using this test, show that this test can correctly evaluate the performance of mortars containing Portland cement or cement mixed with SCMs or combination of SCMs (Stephens & Carrasquillo, 2000). In addition, contrary to ASTM C452-15 (2015), the results obtained in lab with ASTM C1012-18 test procedure (2018) can follow the same trends as in field conditions. Moreover, this method has the advantage to be internationally recognized and broadly applied.

However, the long period of time needed (between 6 and 18 months), in order to reach the final stage of the test, as well as the high concentration of the sulfate solution used, make ASTM C1012-18 (2018) hard to apply and recommend as a routine accelerated test.

The test procedure is summarized in Table 4.2.

4.4.2.3 USBR 4908

ASTM C452-15 (2015) and ASTM C1012-18 (2018) test methods evaluate the resistance of cements to sulfate attack. The performance of concrete is not taken into consideration in these two methods.

The **USBR 4908-86** test method (1986) for length change of hardened **concrete** exposed to alkali sulfate test represents an alternative by examining concrete specimens attacked by sulfate. The test conditions of USBR 4908-86 are similar to the ASTM C1012-18 (2018) ones. The main difference stands in the type of specimens which have a cylindrical shape. The mass variation of the cylinders is measured in addition to the expansion.

The concrete used in the USBR 4908-86 test method can be made with Portland cement or cement with SCMs. The procedure includes three methods where the type of immersion varies as follows:

- full immersion of the cylinders in a 2.1%-Na_2SO_4 solution,
- full immersion in a 10%-Na_2SO_4 solution,
- wetting and drying cycles, where the cylinders are immerged for 16 hours in a 2.1%-Na_2SO_4 solution and then dried for 8 hours at 54 °C.

However, the time needed to reach severe deterioration in this case is estimated at around 2 years and makes the process not really accelerated (Messad, 2009).

4.4.2.4 Canadian Standards (CSA A3004-C6 and CSA A3004-C8)

Portland cements may be evaluated with regard to ESA in Canada by CSA A3004-C6 "Test Method for Determination of Mortar Bar Expansion Due to External Sulphate" (2008) or ASTM C452-15 (2015).

CSA A3004-C8 "Test Method for Determination of Sulphate Resistance of Mortar Bars Exposed to Sulphate Solution" (2008) is used in Canada, in order to assess the resistance of **blended** cements (or blends of SCMs with ordinary Portland cement).

In CSA A3004-C8 (2008), the length change of mortar bars immersed in sulfate solution is the indicator. This method is very similar to the ASTM C1012-18 method (2018). Testing lasts 1 year and can be extended to 18 months. However, CSA A3004-C8 differs by considering two different procedures:

- the first procedure is the same as found in ASTM C1012-18 (2018), where mortar prisms ($25 \times 25 \times 250$ mm^3) are immersed in a Na_2SO_4 solution (with a concentration of 50 g/L) at **23 °C**. This procedure is dedicated to the **ettringite** form of ESA,
- the second procedure applies the same conditions (with regard to mix proportions, casting, molding and curing) as in the previous one, but with a different storage temperature (**5 °C**). This procedure is thus dedicated to the specific effects induced by the formation of **thaumasite**. In this case, mortar prisms are

stored in a refrigerator for 24 hours at 5 °C before being fully immersed in the 50-g/L Na_2SO_4 solution.

The same specifications with regard to C_3A content or expansion as used with ASTM tests are used with CSA A3004-C8 (2008).

4.4.3 The Swiss Test Method SIA 262/1

Usual ESA test methods are based on the principle of immersion where samples are subjected to diffusion of sulfate ions during the attack. This type of tests usually allows one to obtain the change in length and mass parameters, in parallel with visual inspection (Leemann & Loser, 2012). The long duration that takes up to several months and even to 1 year and more can be an important declining factor in reaching fast results, while the concrete and construction industry needs a rapid method in order to assess the performance of concrete mixtures subjected to ESA. Such a rapid test could also be used in controlling concrete production.

In connection with the two “Alp transit tunnel” projects in Switzerland, a test using **drying-immersion cycles** to accelerate sulfate ingress was developed and standardized (SIA 262/1, 2019).

The relevance and efficiency of this accelerated test were studied extensively in lab (Leemann & Loser, 2012; Loser & Leemann, 2016). During this experimental work, the expansion of concrete test specimens with various mix-designs, including five different cement types, exposed to sulfate solutions of different composition was measured. Three water-to-cement (W/C) ratios were used in the test, with five different types of cement. Scanning electron microscopy (SEM) combined with energy dispersive X-ray spectroscopy (EDS) was used to analyze the pre-existing and resulting phase assemblages (e.g., sulfur content) (Leemann & Loser, 2012; Loser & Leemann, 2016). In (Loser & Leemann, 2016), five different cements were used, including Portland cement and industrially-mixed slag cement containing 20–34% OPC and 66–80% blast furnace slag considered as cement with high sulfate resistance (CEM III/B, according to EN 197-1: 2012). The W/C ratios varied between 0.45 and 0.60, resulting in 22 different concrete mixtures. The maximum grain size of the alluvial gravel aggregates was 22 mm.

(150 × 150 × 150 mm^3) cubic specimens were cast and stored at T = 20 °C and RH = 90% for 24 hours. The specimens were then demolded and stored again in the same climate. After 26 days, seven cores with the diameter of 28 mm und a length of 148 mm were taken from a cube of each mixture. This step was followed by applying a plug gauge for strain measurements that was glued at the ends of the test specimens.

The sulfate resistance, for the different specimens and mixtures, was evaluated starting at an age of 28 days on the basis of the Swiss standard (SIA 262/1, 2019). What makes this Swiss standard test method different from what is generally encountered in the literature is the **drying step** before immersion in sulfate solution,

Table 4.3 Summary of the Swiss test method SIA 262/1: 2019 (SN 505262/1) used for ESA testing (based on wetting-drying cycles) (Leemann & Loser, 2012; Loser & Leemann, 2016; SIA 262/1, 2019)

Specimen geometry and size (mm)/Mix proportion	Preparation/Storage before final immersion	Technique/ Measurement	Time required
Core taken from 150 × 150 × 150 mm³ cubes with diam. 28 and length 148 mm Concrete mix with maximum aggregate size of 22 mm	- cubes cast at 20 °C and 90% RH for 24 hours, then demolded in same atmosphere—cores taken on day 26 and stored at 20 °C and 90% RH till day 28 - 5 days of dry condition at 50 °C followed by 2 days of immersion at 20 °C in **50-g/L sodium sulfate solution** (i.e., 34 g SO_4^{2-}/L)	- gauge for strain measurement glued at the ends of specimen - length change + expansion measurement	- curing: 28 days - 4 drying-immersion cycles followed by 8-week immersion (total: **12 weeks**)
Expansion level [‰]	Reproducibility [‰]		
0.2	0.05		
0.77	0.22		
1.79	0.61		

where specimens are put in an oven at 50 °C for 120 ± 2 hours, followed by an immersion in sulfate solution for 48 ± 2 hours. The same drying-wetting cycle is achieved four times in the same manner. Then, the specimens are stored in a 5% Na_2SO_4 solution (i.e., 34 g SO_4^{2-}/L) for 8 weeks. This additional 8-week immersion may yield a significant length change that is noted on a weekly basis. The overall exposure period lasts **12 weeks**, but only the length change during the immersion period is considered for assessing the sulfate resistance.

After the drying-immersion cycles and after the final immersion period, cores were prepared for microstructural analysis by means of environmental SEM. The results of this study showed that the test method enables a rapid assessment of the sulfate resistance of concrete. In addition, the sulfur content in the paste and the formed mineral assemblages were comparable to the ones observed in on-site concrete exposed to ESA.

From then, in Switzerland the sulfate resistance is thus assessed according to Annex D of the Swiss standard SIA 262/1: 2019 (SN 505262/1) (2019). The test procedure is summarized in Table 4.3. The repeatability of the test method is 0.02‰ length change, while the reproducibility depends on the expansion level (see Table 4.3).

In the Swiss standard test method, the criterion for ESA resistance is a maximum relative change in length after 8-week immersion of **0.10%** (SIA 262/1, 2019). It has to be pointed out that a limit value of **0.10%** was defined, since at this length change a decrease in the dynamic E-modulus of the samples is typically observed, indicating damage and crack formation (Loser & Leemann, 2016).

4.4.4 German Common Test Methods

In Germany, the **Wittekindt method**, which dates from 1960, has the longest tradition (Wittekindt, 1960). This method has been used for many years in slightly modified form at the Research Institute of the VDZ (German Cement Works Association). 10 × 40 × 160 mm³ flat mortar prisms are made from standard sand in accordance with the German cement standard DIN 1164 of 1958 with a W/C ratio of 0.60. These are first stored underwater for 14 days at 20 °C, and then—at least three prisms—are stored in sodium sulfate solution (29.8 g SO_4^{2-}/L) and—at least three additional prisms—are stored in water (reference solution) also at 20 °C. The criterion for high sulfate-resisting (SR) cements is a maximum relative change in length after 56-day exposure of **0.5 mm/m**. The relative change in length is taken to be the difference in the change of length between sulfate storage and water storage.

Of greater importance is the **SVA test method** (DIBt, n.d.). It is the method adopted by the committee of experts (SVA) of the German Institute for Construction Technology (DIBt) and which is similar to the Wittekindt method. It also requires flat mortar prisms with the size of 10 × 40 × 160 mm³ (see Fig. 4.19 (Lipus, 2020)). The mortar must be in accordance with European standard EN 196-1: 2016, with a water/cement ratio of 0.50. In contrast to the Wittekindt method, a saturated calcium

Fig. 4.19 Flat mortar prisms used in the SVA test method (Lipus, 2020)

hydroxide solution is used for the preliminary storage and for the reference storage. The test solution is a sodium sulfate solution (29.8 g SO_4^{2-}/L) as for the Wittekindt test method. The criterion for high sulfate resistance is more complex, because the tests have to be performed at 20 °C and at 5 °C. Therefore, the maximum relative change in length must not exceed **0.5 mm/m**, after 91 days at both temperatures and not exceed **0.8 mm/m**, after 182 days for the 20 °C test. In addition to the change in length measurements, the **dynamic modulus of elasticity** must also be measured (DIBt, n.d.). The relative change of this parameter is also taken into account by the committee of experts (SVA) in the assessment. However, no fixed limit values are defined for this purpose. The test procedure is summarized in Table 4.2.

Based on positive results from the SVA test method, cements receive a (general building inspectorate) approval for the use as SR cements. Most of these cements are CEM III/A. Test results of relative change in length of different cements are shown in Figs. 4.20 and 4.21.

In recent years it has been repeatedly shown that the length changes of CEM I-SR specimens can exceed the limit values of the test criterion for the change in length. Also, CEM III/B-SR reference cement samples showed some unexpected behavior when tested according to the SVA test method, e.g., a significant decrease

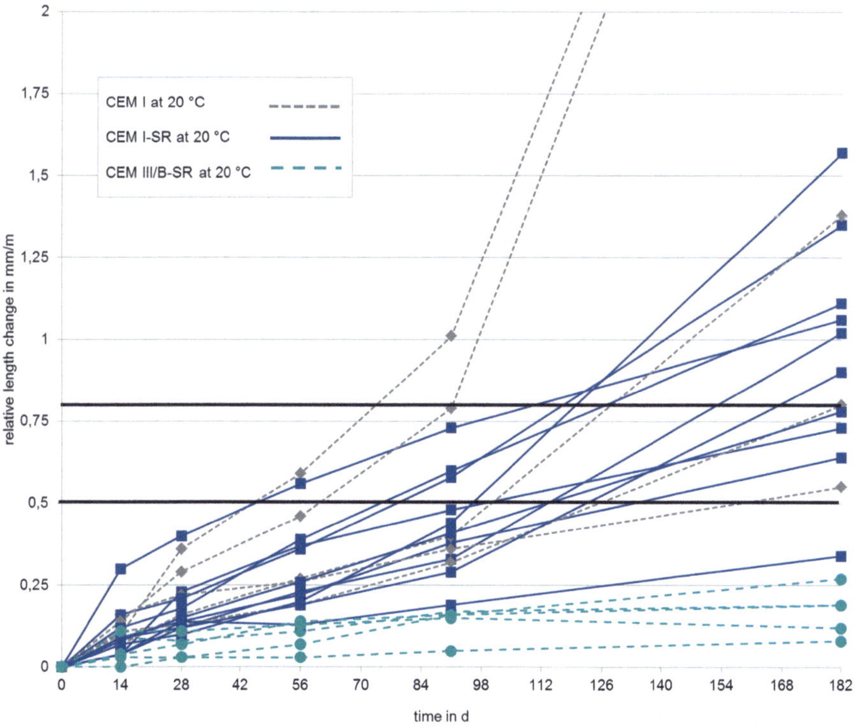

Fig. 4.20 Relative length change according to the SVA test method at 20 °C. Test criteria: 0.50 mm/m after 91 days and 0.80 mm/m after 182 days (Lipus, 2020)

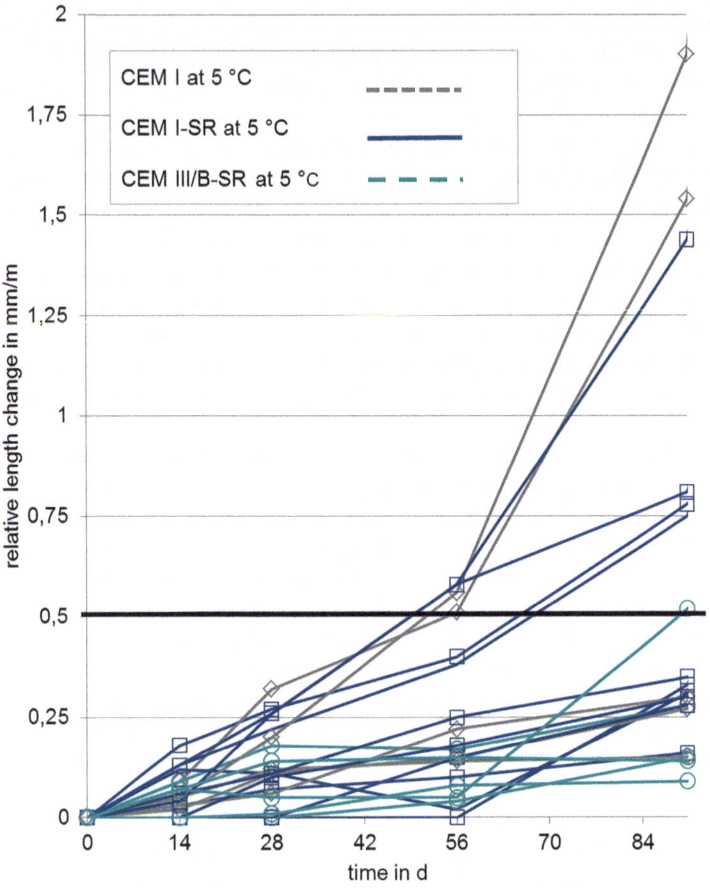

Fig. 4.21 Relative length change according to the SVA test method at 5 °C. Test criteria: 0.50 mm/m after 91 days (Lipus, 2020)

in the dynamic modulus of elasticity. Without doubt, both CEM I-SR and CEM III/B-SR cements have been tried-and-tested cements for concrete with high sulfate resistance for many years. The fact that this situation is not always reflected in the lab by the SVA test method illustrates that the method has some weaknesses in the assessment of the SR property for practical application. A VDZ research project has shown that this is mainly due to the formation of secondary gypsum, which can form in CEM I-SR cements in the high sulfate concentrations used in the SVA test method (VDZ Report, 2017).

Since the SVA test method was apparently too strict, the committee of experts (SVA) proposed a modification and the DIBt changed the test conditions, as mentioned, e.g., in (VDZ Report, 2017; Haufe & Vollpracht, 2019; Lipus, 2020). The concentration of sulfates was reduced to 3 g/L (0.44%-Na_2SO_4 solution). As a matter of fact, this value is more realistic and the expansion at this lower concentration is primarily caused by ettringite rather than by gypsum. However, due to lower degradation, the test then takes at least 2 years, and since no limit value exists for these conditions, a CEM I-SR and a CEM III/B-SR have to be tested in parallel, in order to compare the performances. At the time being, the normal SVA test method is only used for CEM-III cements. All other binders have to be tested with the **modified SVA test method**, using a sulfate concentration of 3 g/L (Lipus, 2020).

4.5 Improved Methods Recently Developed for Sulfate Resistance Testing

Examples of improved test methods recently developed are detailed in the following sections. The methods described allow the assessment of ESA resistance in full immersion or in wetting-drying cycle conditions.

4.5.1 The French Approach

4.5.1.1 Review of the Various Test Methods Initially Available

Within the framework of the development and implementation of comparative (performance-based) durability approaches (CEN/TR 16563, 2013; Linger et al., 2014) various research works and projects have been carried out in France, or were still in progress when this report was drafted, in order to design relevant (accelerated) lab test methods for the assessment of the ESA resistance of concretes (see, e.g., Rozière, 2007; Messad, 2009; PerfDub Project, 2014–2019; Massaad, 2016; Jabbour, 2019). The experimental work was developed, in order to take into account as far as possible the on-site conditions.

The French tests described in Table 4.4 have been designed from the analysis of the mechanisms of degradation and from existing test methods, which can be found in the state-of-the art sections of (Rozière, 2007) and (CEN/TR 15697, 2008). The common parameters of the three tests described in the following (see Table 4.4) are:

- pH control, which yields better representativeness of real exposure conditions, as well as shorter tests,
- relatively low sulfate concentration.

The selected values of these parameters are in agreement with the exposure conditions listed in the European Standard EN 206: 2013.

The **sulfate concentration** in the test procedure proposed initially (3 g/L, diffusion, concrete) can be regarded as long-term testing, rather dedicated to obtaining reference data. *Messad* has chosen a higher sulfate concentration (6 g/L) (Messad, 2009; Messad et al., 2010) but which is within the range considered in EN-206: 2013 exposure classes.

Table 4.4 Test conditions of French tests (Rozière et al., 2015)

Exposure	Immersion		
pH	pH = 7.5 controlled by adding acid		
Sulfate concentration	Parameter of experimental study/recommended values: XA2: 4.4 g/L Na_2SO_4 (3 g/L SO_4^{2-}) XA3: 8.9 g/L Na_2SO_4 (6 g/L SO_4^{2-})		
Age of materials at exposure	Parameter of experimental study Recommended value: 28 days		
Use of the test	1. Reference data 2. Performance testing	Study of degradation mechanism	Comparison of mixtures: accelerated test (3 months)
References	*Roziere's* work (Rozière, 2007; Rozière et al., 2009)	*El Hachem's* work (El Hachem, 2010; El Hachem et al., 2012a, b; Massaad et al., 2015)	*Messad's* work (Messad, 2009; Messad et al., 2010)
Specimens	1. Concrete: 70 × 70 × 280 mm^3 2. Mortar: 20 × 20 × 160 mm^3	Mortar - 20 × 20 × 160 mm^3 (El Hachem, 2010) - cyl. 20 × 160 mm (Massaad et al., 2015)	Concrete - Expansion: 70 × 70 × 280 mm^3 - Indicators: cyl. 110 × 220 mm
Curing and preconditioning	Water curing, 28 days	Parameter of experimental study Recommended: Water curing, 28 days	Water curing, 28 days 60 °C drying **Saturation under vacuum by sodium sulfate solution**
Temperature	20 °C	20 °C	25 °C
Solution renewed every	/ 4 weeks	15 mL of added acid	/ 4 weeks
Duration of test	1. 9 years 2. 3–24 months	3–24 months	Parameter of experimental study Recommended: 3 months
Setup (including pH control)	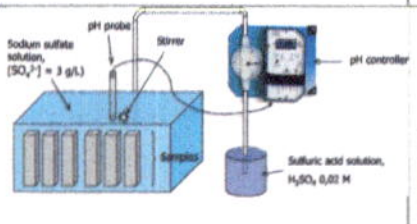	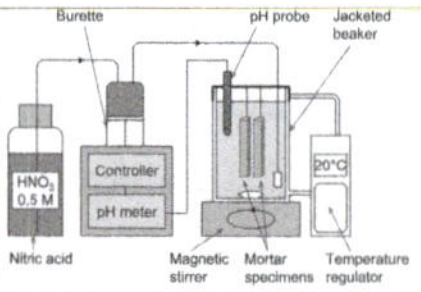	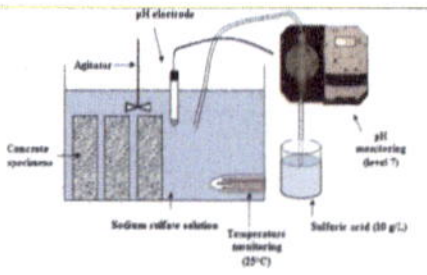

In (Messad, 2009), the ingress of sulfate ions is accelerated through **pre-drying** of the specimens at T = 60 °C (followed by saturation under vacuum).

Depending on the test method applied, **mortar** or **concrete** samples are used (see Table 4.4). The use of mortar as tested material results in shorter testing time, since specimens are smaller. However the link between mortar behavior and concrete behavior is still not well understood (see Sect. 4.2.5). An experimental study has been done on this topic in (Massaad, 2016).

The **size of the specimens** is selected according to the tested material (on the basis of the maximum aggregate size). When mortar samples were tested, since the wider the section the longer the testing time, the chosen dimensions were 20 mm in width or diameter and 160 mm in length. This is consistent with ASTM C1012-18 standard (2018), which recommends 25 × 25 × 285 mm^3 specimens.

Moreover, if the study deals with sulfate resistance of different binders, a 0.6 water/binder ratio can be chosen (CEN/TR 15697, 2008; El Hachem, 2010; Massaad et al., 2015).

The test procedure proposed initially has been improved during *El Hachem's* PhD thesis (El Hachem, 2010). The pH was automatically controlled and the sulfate solution renewed every 15 mL of added acid and analyzed. This allows a better control of the test conditions, as well as the quantification of the fluxes of hydroxide, calcium, and sulfate ions, in order to better investigate the degradation mechanisms.

The approach proposed in *Massaad's* PhD thesis (Massaad, 2016) is based on four parameters, which are assessed experimentally: expansion, mass (saturated, surface dried specimens), hydrostatic weighing, and leached hydroxide amount. This approach allows a better monitoring of the test and the assessment of parameters likely to describe the different phenomena involved in the ESA process in lab, such as leaching and ettringite/gypsum precipitation.

4.5.1.2 LMDC-GeM Improved Accelerated Test Method

Description of the Test Method

The LMDC-GeM improved accelerated test method is based on *Messad's* work (Messad, 2009; Messad et al., 2010). The following approach is applied, in order to accelerate the attack:

- monitoring of the *temperature* of the solution (set at 25 °C),
- use of high sodium sulfate concentration: 8.9 g/L (i.e., 6 g/L SO_4^{2-}). This is the highest sulfate concentration mentioned in EN 206: 2013 (**XA3** class),
- *continuous pH control* of the solution (set at 7) during the different periods of the test by adding sulfuric acid,
- the pore volume of the samples is *pre-saturated* (by hydrostatic weighing) by a sodium sulfate solution. This step constitutes the originality of the accelerated test, compared to other tests encountered in the literature, and is an essential factor in the acceleration of the attack.

After 28-day wet curing (and before the immersion test itself), the concrete samples are preconditioned as follows: they are oven dried at T = 60 °C and then saturated under vacuum for 48 hours (NF P18-459, 2022) by a Na_2SO_4 solution at a concentration of 8.9 g/L. The duration of the whole experimental work is approximately **17 weeks** (4-week curing +1-week preconditioning +12-week contact with sulfates).

The test conditions are summarized in Table 4.5 and the setup used is depicted in Fig. 4.22 (Messad, 2009).

Various parameters are measured before (on sound material), during and after the 12-week immersion period (on degraded material):

Table 4.5 Conditions of the LMDC-GeM improved accelerated test method

Preconditioning	- drying phase of the concrete specimens at 60 °C until constant mass - 48-hour saturation under vacuum with 8.9 g/L sodium sulfate solution
Exposure conditions	Full immersion
pH level	pH = 7—Controlled by sulfuric acid addition
Sulfate concentration	8.9 g/L Na_2SO_4 (i.e., 6 g/L SO_4^{2-})
Temperature	T = 25 °C—Controlled
Renewal of the solution	Every 4 weeks
Duration of the exposure	12 weeks

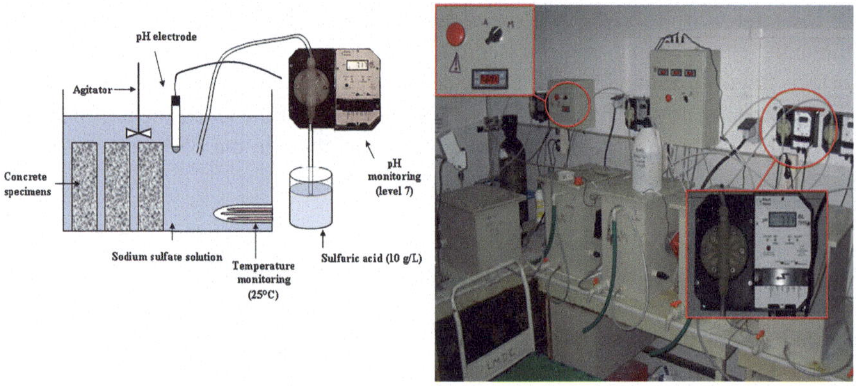

Fig. 4.22 Illustration of the setup used for the LMDC-GeM improved accelerated test method (Messad, 2009; Linger et al., 2014)

Table 4.6 Concrete properties measured on sound and "degraded" concrete

Test	Test method
Compressive strength (MPa)	EN 12390-3: 2019
Chloride diffusion coefficient (10^{-12} m/s)	non-steady-state migration test according to NF XP P18-462 (2022)
Water porosity (%)	NF P18-459 (2022)

- *Longitudinal expansion and mass variation*: measured on concrete prism specimens (70 × 70 × 280 mm^3) every 2 weeks over a period of 12 weeks. Each measurement is made on three specimens,
- *Materials properties:* Table 4.6 summarizes the tests applied on concrete cylindrical specimens (diam. 110 × 220 mm), in order to quantify the effect of ESA on the concrete properties (measurement applied before and after the 12-week immersion test).

Examples of Results Obtained with the Described Test Method

Figure 4.23 presents longitudinal expansion and chloride diffusion coefficients results obtained on 8 concretes with various binders (100% OPC for reference, 10%, 20%, 30% fly ash (V), and 10%, 20%, 30%, 45% slag (S)). The effect of the

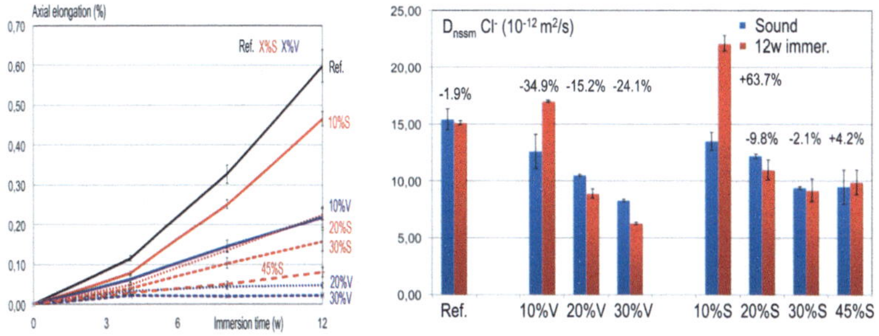

a) Longitudinal expansion

b) chloride diffusion coefficient obtained from non-steady-state migration test (D_{nssm}) on sound and "degraded" concrete

Fig. 4.23 Effect of the SCM on ESA as quantified by the described test method—Variation of properties of concretes degraded by ESA (Messad, 2009). (**a**) Longitudinal expansion. (**b**) chloride diffusion coefficient obtained from non-steady-state migration test (D_{nssm}) on sound and "degraded" concrete

SCM nature and substitution ratio on ESA as quantified by the described test method is clearly pointed out.

4.5.1.3 IFSTTAR Test Method

The experimental setup of the LMDC-GeM accelerated test method described in the previous section has been further improved/optimised in various labs in France.

For example, the automated setup developed at IFSTTAR includes a large tank with a capacity of approximately 80 L, acting as reservoir of sulfate solution for multiple simultaneous tests in various possible configurations (e.g., full immersion, partial immersion, wetting-drying cycles, etc. see Fig. 4.24). The sulfate solution used here has a concentration of 15 g/L. This concentration is slightly higher than the one adopted in (Messad, 2009), in order to detect the potential degradation earlier but without changing the mechanisms. A pump with a maximum flow rate of 750 L/h is immersed in this large tank, in order to transfer the sulfate-bearing solution into several smaller tanks (six), where the volume is adjusted to 10 L in each one in the case of semi-immersion (see Fig. 4.24). Upstream of these tanks, the solution crosses a cylindrical tube where the pH of the solution is constantly monitored and regulated to 8.0 ± 0.1 by means of a continuous titration system with sulfuric acid H_2SO_4 (0.02 M, i.e., 0.05%) before it flows back into the reservoir tank. The added sulfuric acid is properly mixed in the sulfate solution as this one is continuously moving between the sulfate reservoir and the specimen tanks (process 3, see Fig. 4.24). In order to keep the concentration of the solution in the baths

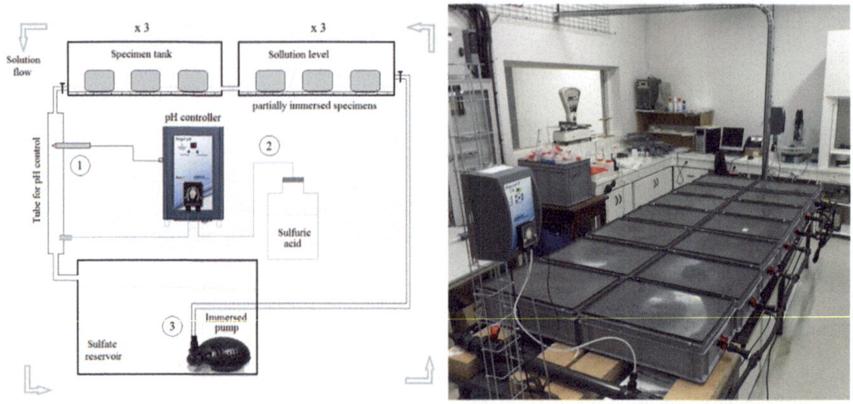

a) Scheme of the system b) picture of the setup

Fig. 4.24 Scheme (**a**) and picture (**b**) of the setup dedicated to ESA accelerated test developed at IFSTTAR, which includes pH-control and multiple simultaneous testing in various configurations (Jabbour, 2019)

constant, the solutions are renewed each week to ensure an unlimited reservoir of sulfate ions. In addition, this renewal makes it possible to provide an infinite source of sulfate ions, as this is the case in real exposure conditions.

Here, both prisms ($70 \times 70 \times 280$ mm^3) and cylinders with 110 mm diameter and 220 mm length are used to carry out the measurements.

4.5.1.4 Input from French National Project *PerfDub* (2014–2019)

The French National Project *PerfDub (Performance-based approach of durability of concrete structures)* was launched at the end of 2014 and regroups several French labs (PerfDub Project, 2014–2019). This project aimed to define a methodology for the implementation of performance-based approaches. More precisely, the Working Group "External chemical attacks", which includes in particular ESA aspects, was working on the development of a procedure for accelerated ESA testing and carried out round robin tests on a broad range of concrete mixtures (IREX, 2023).

Various aspects were studied within this framework, such as:

- geometry of the sample,
- impact of incorporation of SCMs,
- comparison between test methods: the previously developed French protocol based on *Messad's* work and involving (saturation + full immersion) (Messad, 2009; Messad et al., 2010) was compared to the Swiss test method SIA 262-1 (2019) detailed in Sect. 4.4.3, involving drying-immersion cycles,
- assessment of kinetic coefficient based on real cases,
- testing on already degraded concretes.

At the end of the *PerfDub* Project (IREX, 2023), it was concluded that the **test method based on *Messad's* work** (Messad, 2009), with:

- moulded prisms (70 × 70 × 280 mm^3),
- 8.9 g/L-Na_2SO_4 solution (i.e., 6 g/L SO_4^{2-}) (**XA3**),
- preconditioning +12-week immersion (in the solution),
- pH = 7 regulation,
- T = 25 °C,

where the following measurements were carried out:

- weight and length changes (on prisms),
- evolution of compressive strength, water porosity and chloride diffusion coefficient (on cylinders),

provided as good results as those obtained with the **Swiss test method SIA 262-1** (2019) described in Sect. 4.4.3, for which experience has been acquired for a long time, and with:

- drilled cores (diameter 28 mm),
- Na_2SO_4 solution with 34 g/L SO_4^{2-},
- variable pH,
- T = 20 °C,
- test duration: 12 weeks (4 drying-immersion cycles +8-week immersion),

despite different preconditioning, exposure conditions and sample types.

Consequently, **both methods could be recommended**, and the choice between them will be based on the actual field exposure conditions.

Within the framework of the *PerfDub* Project, in the case of **XA3 exposure class** (highly aggressive chemical environment), a limit value of:

- 0.5‰ for expansion after 12-week immersion (test method based on *Messad's* work (Messad, 2009)),
- 1.2‰ for expansion after the 8-week full immersion period in the case of the Swiss test method (see also Sect. 4.5.3),

has been selected (IREX, 2023).

4.5.2 *The Belgian Approach: Accelerated Test by Alternating Wetting and Drying Cycles*

In the Belgian test method described in (Gruyaert et al., 2012), the accelerated degradation of concrete specimens exposed to a sulfate solution is based on **wetting and drying cycles**. **Cylindrical concrete samples** are used in this case and expansion is measured by using **laser sensors**.

Fig. 4.25 Apparatus for accelerated sulfate resistance testing on concrete specimens (Gruyaert et al., 2012)

The samples are subjected to a continuous rotation and turned alternately through the sulfate solution and the air (see Figs. 4.25 and 4.26). The apparatus was also used to compare the acid and sulfate resistance (Gruyaert et al., 2012). The distance between the samples and the laser sensors is set at 0.1 mm. The resistance to ESA is investigated by using a contact solution with high sulfate concentration (29.8 g/L Na_2SO_4 and 29.8 g/L $MgSO_4$). The duration of the test is **14 days**. The solution is changed on a weekly basis into a fresh one, in order to keep the conditions of the attack appropriate.

Before the beginning of the cycles, all samples are properly brushed, in order to eliminate any type of errors or irregularities that might happen during the test. The dimensions are also measured by means of laser sensors. Since contact with sulfates will induce significant expansion, the change in radius was measured before, as well as after brushing. The change in radius vs. time is monitored (see the results displayed in Figs. 4.27 and 4.28).

Note that in such a test, thenardite (see Sects. 4.1.1, 4.2.3, and 4.7.1) and/or calcite could form in the outermost concrete layers (Jabbour, 2019; Henrique de Bem & Medeiros, 2020). The buildup of salt crystallization pressure in the parts which are not submerged may yield poor performance of concrete containing high amounts of GGBS (e.g., 85%) (Gruyaert et al., 2012).

4.5.3 The Swiss Approach

The protocol of the **Swiss test method SIA 262/1:2019** (2019) detailed in Sect. 4.4.3 can be summarized as follows:

- drilled cores (diameter 28 mm)
- Na_2SO_4 solution with 34 g/L SO_4^{2-}
- variable pH

Fig. 4.26 Image of concrete test specimens before and after 37 weeks of cyclical exposure to Na_2SO_4 solutions (Gruyaert et al., 2012)

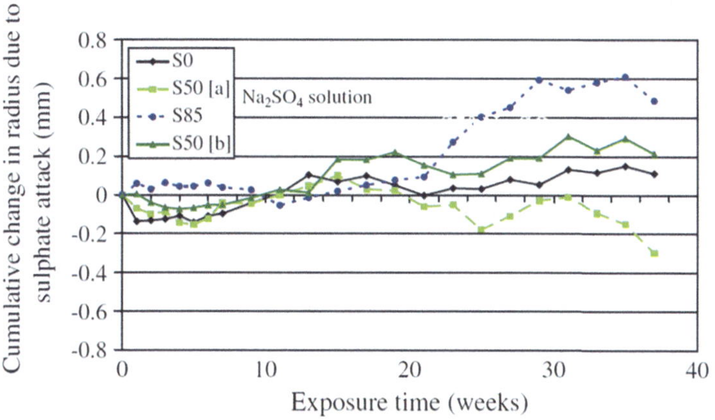

Fig. 4.27 Cumulative change in radius due to sulfate attack for test specimens cyclically exposed to a 29.8 g/L Na_2SO_4 solution (S0, S50[a] and S85) or a 3 g/L Na_2SO_4 solution (S50[b]) (Gruyaert et al., 2012). *Mixes with 0 mass% (S0), 50 mass% (S50) and 85 mass% (S85) of the cement replaced by BFS*

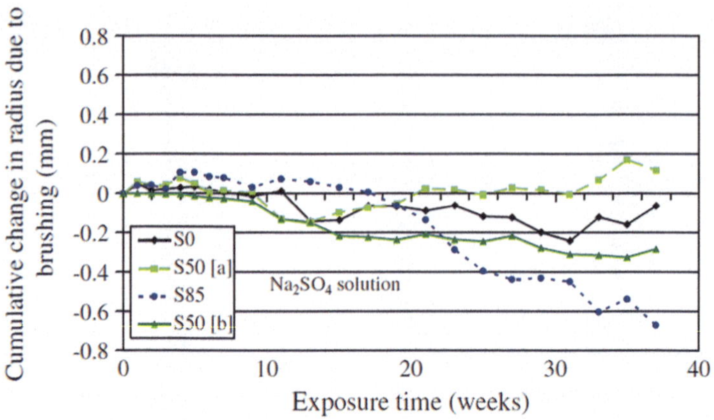

Fig. 4.28 Cumulative change in radius due to brushing for test specimens cyclically exposed to a 29.8 g/L Na_2SO_4 solution (S0, S50[a] and S85) or a 3 g/L Na_2SO_4 solution (S50[b]) (Gruyaert et al., 2012). *Mixes with 0 mass% (S0), 50 mass% (S50) and 85 mass% (S85) of the cement replaced by BFS*

- T = 20 °C
- test duration: 12 weeks (4 drying-immersion cycles +8-week immersion)

A variation of this test method was applied in 2016 and 2017 in connection with the bidding process for tubbing segments in a new railway tunnel (Bözberg) (Hartmann et al., 2013; Loser & Leemann, 2016). The groundwater in the tunnel has been studied in detail and is known to contain sulfates with a concentration of up to 2 g/L in the area of the planned tunnel (Hartmann et al., 2013). In order to prove the suitability for this environment, concrete mixtures of several applicants were tested in the following way:

- *Variation 1*: The same protocol as in the Swiss method was used for the drying-immersion cycles, but the following immersion period was prolonged from 8 to 16 weeks. The limit value for expansion at immersion time of 8 weeks was set to **1.2‰** instead of 1.0‰. The expansion at immersion time of 16 weeks was only used for informative reasons.
- *Variation 2*: The same protocol as in the Swiss method was used for the drying-immersion cycles. The following immersion period was conducted at **7 °C** instead of 20 °C and its duration was **16 weeks**. The **dynamic E-modulus** during the immersion period had to be determined in addition to the length and mass changes (Loser & Leemann, 2016). Moreover, it was required to study the concrete at the end of the test using scanning electron microscopy (SEM) combined with EDS. The **limit value for expansion was 1.2‰** after an immersion of 16 weeks and a **reduction of the dynamic E-modulus less than 10%**.

Experience has shown that in general, that the cores expand faster when stored in sulfate solution at 20 °C, but in some cases the final expansion was higher for the cores stored at 7 °C.

4.5.4 The German Approach

4.5.4.1 Preliminary Remark

As described in Sect. 4.4.4, the current modified SVA test method takes at least 2 years, which is quite inconvenient for cement producers. Two different approaches are currently followed to improve this situation. The VDZ recently developed a modified expansion test method in a research project (VDZ Report, 2017; Lipus, 2020). At the same time, a new test method based on tensile strength measurements was developed by the ibac (RWTH) and the FEhS Institute in Duisburg.

4.5.4.2 VDZ Test Method

The modified version of the SVA test method does not yet have extensive experience. But the VDZ test variant yielded promising results. The test solution used has a sulfate concentration of 3 g/L. The solution is a mixture of calcium, magnesium and sodium sulfates, similar to real groundwater compositions. The cylindrical specimen geometry in conjunction with a tumbling motion during the first hours of setting and hardening gives more homogeneous test specimens, which is advantageous for a lower dispersion of the test results (see Fig. 4.29). This test method made it possible to differentiate more clearly between cements with and without SR properties. The test results of the SR cements showed no "false" results. First tests with cements with several main constituents are also rated positively. In contrast to the blastfurnace slag-rich CEM III cements, the CEM II/B-S showed significantly higher expansion. Differences in the expansion behavior of the fly ash-containing cement mixtures could also be observed, which correlate with different levels of sulfate resistance of the cement-fly ash mixtures and are due to the temperature impact (see Fig. 4.30).

The test method provides meaningful results for cements with various main constituents, in particular for cements containing limestone, slag and fly ash (Lipus, 2020). However, the results obtained so far with the **VDZ test method** are to be verified with further tests on more cements. It is also necessary to clarify in more details what the optimal test period is. Also, there are still no comparative experiments and thus no information about comparability of the test results. Likewise, the repeatability of the test method has not been so far extensively studied. All these

Fig. 4.29 Cylinder mortar specimens used in the VDZ test method (Lipus, 2020)

points are still necessary if the method should be considered as an alternative to the previous methods.

4.5.4.3 Performance Test Based on Tensile Strength, Developed by Ibac (RWTH) and FEhS

The idea of using bending or tensile strength as an indicator for sulfate induced degradation dates back to 1960 (Koch & Steinegger, 1960) and 2002 (Mulenga, 2002), respectively. The tensile strength is more sensible to sulfate damage compared to bending strength, however, measuring the tensile strength is associated with higher effort to prepare the specimen and the scatter of testing is quite high.

In order to improve the applicability of tensile strength measurements, *Haufe* et al. tested the use of briquet specimen according to ASTM C307-18 (2018) (Haufe & Vollpracht, 2019). This standard provides a quite simple method for tensile strength testing. There is no need to apply stamps to the end of the specimen. The shape of the samples and the testing device are shown in Fig. 4.31. The thickness of the briquets is 25.4 mm. Load application and geometry have been optimized to achieve relatively low scatter, allowing a more precise detection of damage (Neunzig et al., 2018).

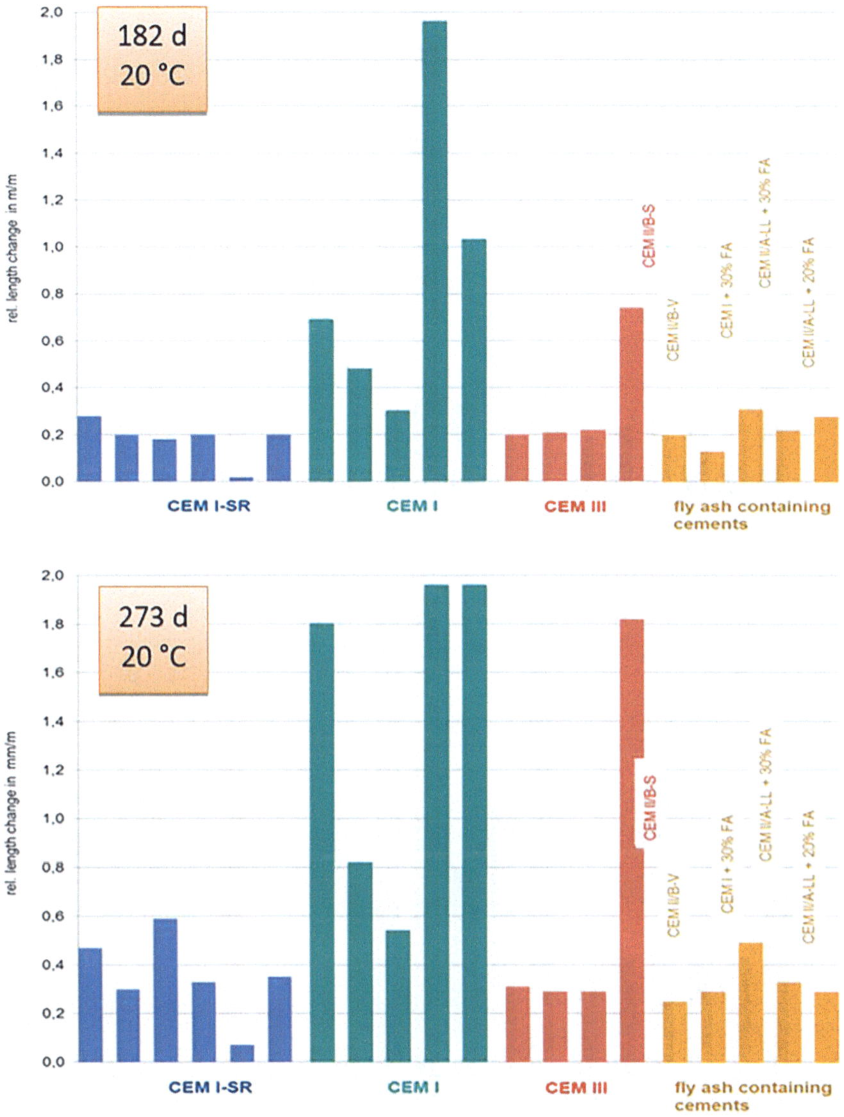

Fig. 4.30 Relative length change results from VDZ test method after 182/273-day storage at 20/5 °C (Lipus, 2020)

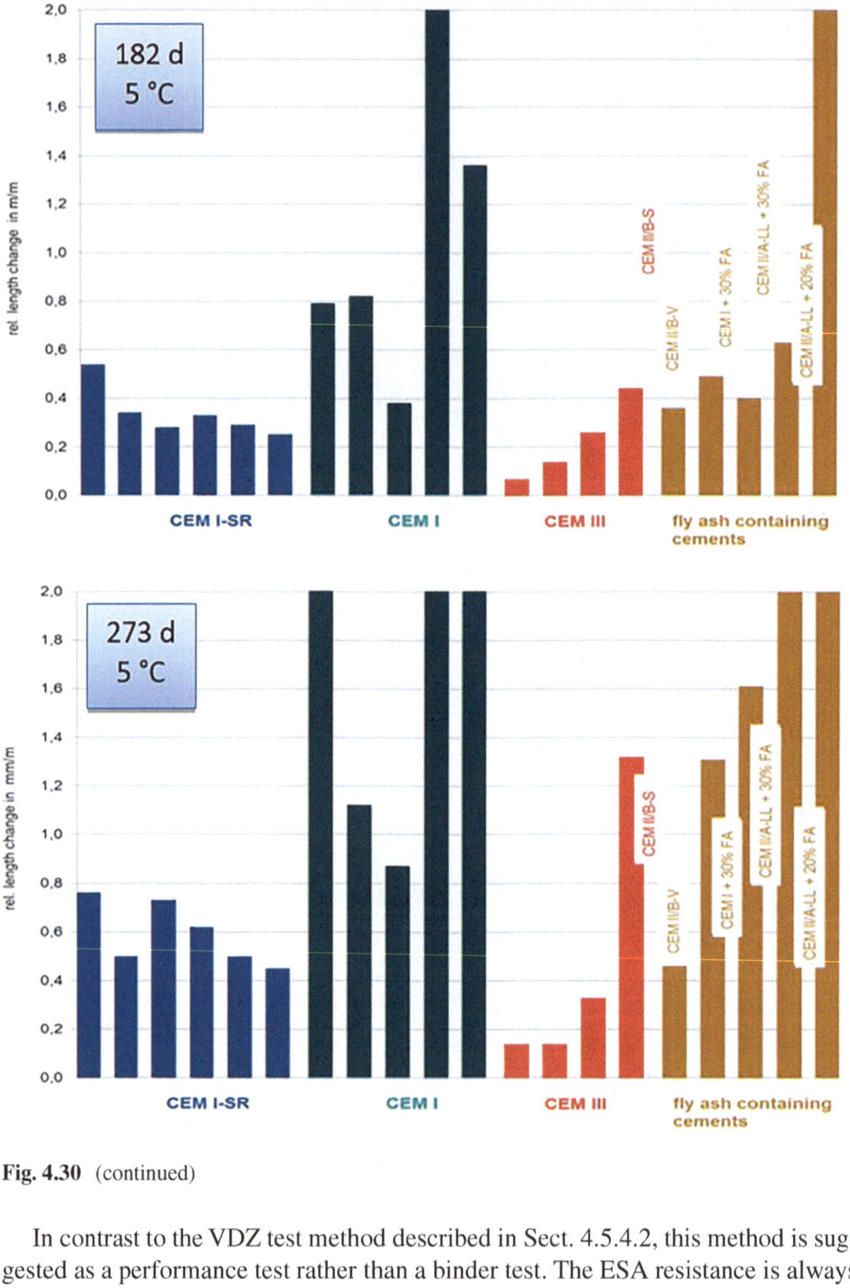

Fig. 4.30 (continued)

In contrast to the VDZ test method described in Sect. 4.5.4.2, this method is suggested as a performance test rather than a binder test. The ESA resistance is always a combination of physical and chemical properties of concrete. A performance test should therefore use concrete specimens with variable mix-designs rather than mortar samples with standardized binder contents and w/b ratios. Hence, the proposed approach suggests determining the loss of tensile strength of briquet concrete samples stored in sulfate solution with realistic concentration.

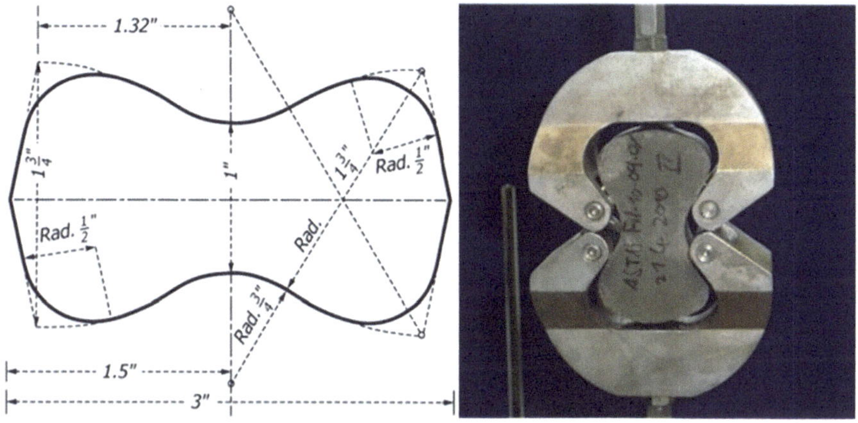

Fig. 4.31 Briquet specimen and tension clips according to ASTM C307-18 (2018) (dimensions given in inch = 25.4 mm)

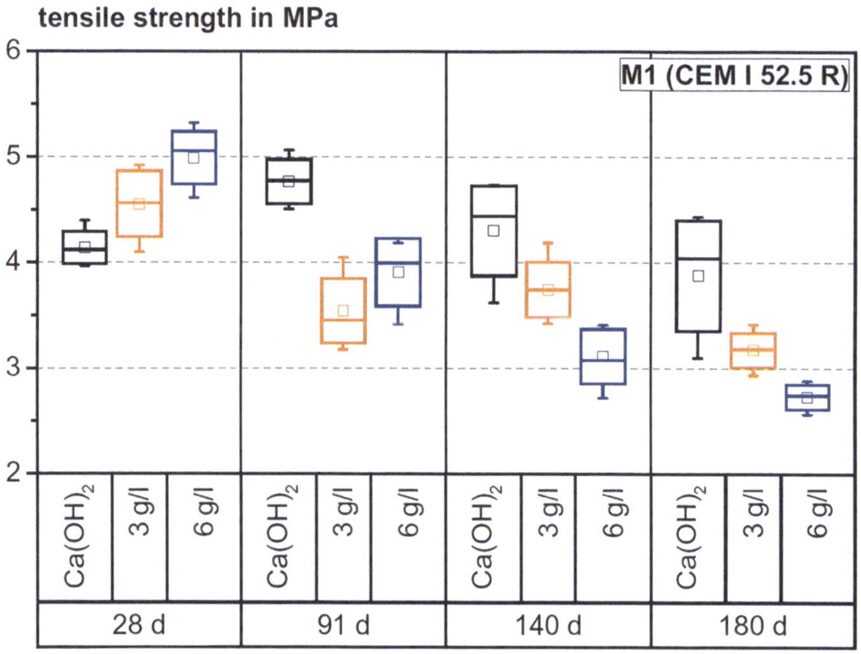

Fig. 4.32 Tensile strength results of a CEM-I concrete with w/c = 0.5 and a maximum grain size of 8 mm, stored in saturated $Ca(OH)_2$-solution (reference), sodium sulfate solution with 3 g SO_4^{2-}/L and sodium sulfate solution with with 6 g SO_4^{2-} /L at 5 °C for 28, 91, 140 and 180 days (Haufe & Vollpracht, 2019)

In (Haufe & Vollpracht, 2019) a first set of concretes was tested at several storage conditions. Some exemplary results for a non sulfate-resistant mix are shown in Fig. 4.32.

During the first 28 days, a strength increase of the samples stored in sulfate solutions compared to the reference storage can be observed. Apparently ettringite was precipitating in the pore system at the beginning of the test and contributed to strength. At later ages however, the ongoing sulfate reactions lead to a significant strength decrease. The sulfate concentration of 6 g/L led to a higher strength loss compared to storage in a 3-g/L sulfate solution and therefore may constitute a reasonable compromise between acceleration and sulfate concentrations found in the field.

Based on this preliminary work, a joint research project of ibac and FEhS was funded by the German Federation of Industrial Research Associations (AiF), in order to optimize testing conditions and to develop testing criteria for this performance test. To implement the test in the process of technical approvals in Germany, a round robin test is needed to determine the reproducibility and reliability of the procedure.

4.5.5 The New Approach in North America

NIST and the Univ. of Toronto have developed a new test for determining the sulfate resistance of cements. The new test consists in measuring the expansion of **small cement paste bars** ($10 \times 10 \times 60$ mm^3) while in a sulfate solution (Ferraris et al., 2018). The background study that led to these dimensions can be found in references (Ferraris et al., 2005; Ferraris et al., 2006; Aleksic & Hooton, 2011). This method was presented to ASTM for standardization, but was not yet adopted, in 2018, when the paper which describes the detailed protocol was published (Ferraris et al., 2018). In 2015, this test method has been part of an interlaboratory study which included also tests on ($25 \times 25 \times 280$ mm^3) prisms as described in ASTM C1012-18 (2018).

The driving force for the development of this new test was to have a determination of the resistance to ESA faster than 1 year, as required by ASTM C1012-18 (2018). The small cement paste bar test provides the same information within 3 months. This test is applicable to Portland cements, blends of Portland cement with pozzolans or slag, and blended hydraulic cements.

4.6 Promising Researches on Original and Advanced Testing and Detection Methods

Innovative methods of acceleration of the processes such as the migration of sulfates under an electrical field as it is usually applied with chlorides, as well as enhanced use of innovative NDT methods to detect ESA-induced damage, have been recently proposed. Examples of innovative testing and detection methods are detailed in the following sections.

4.6.1 Accelerated Methods by Means of Electrical Field

It is well known that sulfate and chloride ions are both anions which make it feasible to use electrical field, in order to increase the rate of ingress of these ions inside cement-based materials.

Based on this (electrochemical polarization) principle, a rapid test has been proposed several decades ago, in order to assess the ability of concrete to resist sulfate penetration (Tumidajski & Turc, 1995). The results correlated well with the sulfate penetration depth determined from 2-year ponding experiments and indicated that the technique was useful for predicting sulfate penetration into concrete.

More recently, the same way to accelerate sulfate ingress in the case of full immersion of mortar or concrete samples, was developed by *Lorente* et al. (Lorente et al., 2011) and by *Huang* et al. (Huang et al., 2015; Huang et al., 2016). In these tests, a so-called "combined" attack is applied (i.e., exposure to sulfate solution in full immersion and to electrical field).

In the experimental protocol proposed in (Huang et al., 2015), prismatic mortar samples (40 × 40 × 160 mm^3) are cast in the middle part of H-shaped molds (see Fig. 4.33a) and the accelerated ESA under an electrical field (see Fig. 4.33b) is applied to them.

Another setup was tested later by the same group of researchers (Huang et al., 2016). In this case, the prismatic mortar specimens (40 × 40 × 60 mm^3) are cast in the middle part of a special mold split into small units (as shown in Fig. 4.34a). 24 hours after casting, the movable parts of the special mold are pulled out and the specimens along with mould are cured for 27 days at 20 °C and 95% RH. Then, the cathodic compartment is filled with a 5%-Na_2SO_4 solution while the anodic compartment is filled with a 2.8%-NaOH solution (see Fig. 4.34b). A low-frequency pulsed electrical field is used in this investigation. The test is carried out according to the following protocol: 30 V-voltage applied for 20 s followed immediately by another 20-s interruption and lasts 90 days (Huang et al., 2016).

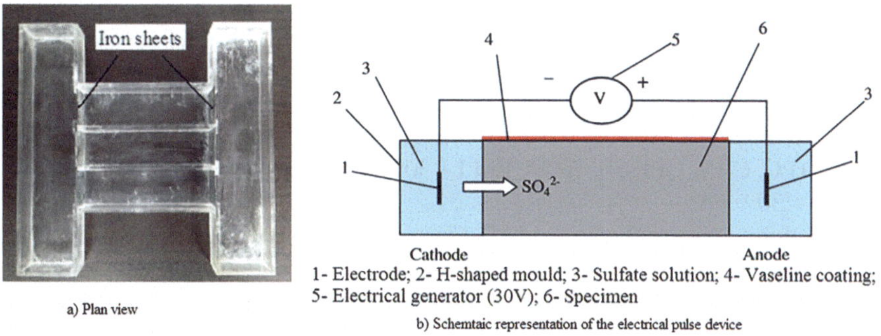

Fig. 4.33 Accelerated ESA test under an electrical field, proposed in (Huang et al., 2015)

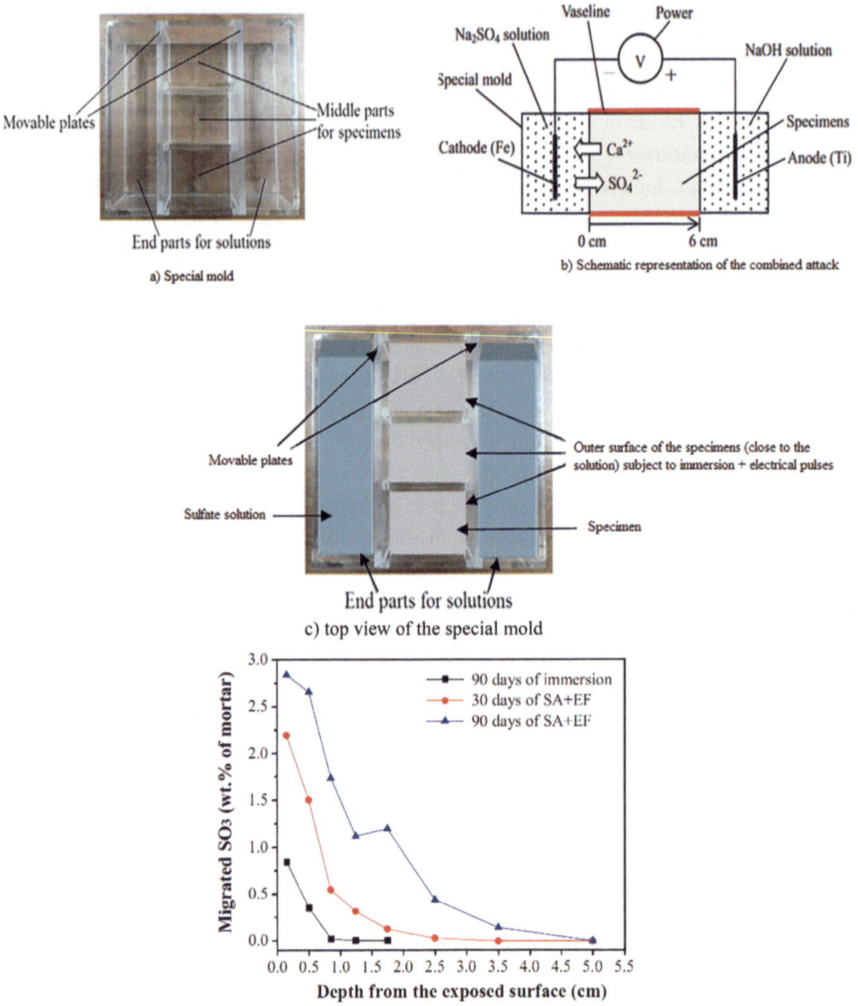

Fig. 4.34 Accelerated ESA test under an electrical field, proposed in (Huang et al., 2016), and example of results. (**a**) Special mold (**b**) Schematic representation of the combined attack (**c**) top view of the special mold (**d**) SO_3 content in PC mortar samples after immersion or migration test, as a function of the depth from the exposed surface

In reference (Huang et al., 2016), various types of binder were used to test the system (PC: CEM I, LP-30: 70% CEM I + 30% limestone powder, and FA-30: 70% CEM I + 30% FA).

In both references (Huang et al., 2015) and (Huang et al., 2016), measurements were carried out, in order to evaluate the consequences of electrical pulses during the sulfate attack. The sulfate concentration corresponding to the SO_3 content in mortar samples was measured. In addition, the mechanical properties of the attacked samples (compression and flexural strengths) were evaluated. To identify the changes in the microstructure and the mineral phases at the end of the accelerated attack, a microscopic observation was performed using scanning electron microscope and X-ray diffraction techniques.

It was observed that accelerating the ESA via electrical pulse technique helped not only in increasing the migration of sulfate ions into the samples but also in accelerating the leaching process which leads to a complete dissolution of portlandite (CH) followed by a decomposition of C-S-H. The overall increase in the amount of sulfate ingressed into the mortar samples (see Fig. 4.34d) induced a severe deterioration and damage by the formation of both ettringite and gypsum. This was confirmed by the mechanical strength losses observed after the attack (Huang et al., 2016).

As a conclusion, such an accelerating technique, if properly applied, can be considered as a fast method, in order to study the effects of ESA on cement-based materials. However, the severe leaching, which occurs during the test (Huang et al., 2016), can appear as somewhat unrealistic. More generally, the nature of the mechanisms and the reaction products formed, or at least the chronology of occurrence of these products, may differ compared to other types of tests and to field conditions (Lorente et al., 2011).

4.6.2 Enhanced Use of NDT: Electrochemical Impedance Spectroscopy (EIS)

Electrochemical impedance spectroscopy (EIS) can be used to investigate ESA in concrete or RC by non-destructive method, on the basis of the measured electrical properties and previously established models for samples exposed to chlorides. For example, according to (Bragança et al., 2014), the formation of gypsum and monosulfate products can be identified in the matrix, in particular during the initial stages of the degradation process, for reinforced mortar specimens exposed for approximately 300 days in a sulfate chamber, by using such a technique.

4.6.3 Combined Monitoring of ESA and Leaching: Measurement of Mass/Length Change and Ionic Concentration

In most of the studies, two stages have been identified during the ESA process from length change measurements (see Sect. 4.2). The first stage is a dormant period where no or little expansion can be measured. Then, significant expansion occurs. As previously mentioned, expansion data do not provide enough information, in order to assess the chemical reactions and to understand the mechanisms causing damage. Additional information is needed in particular to explain the phenomena which occur during the first stage of sulfate attack.

Experimental procedures have hence been designed, which involve the **monitoring of leaching** on mortar or concrete specimens (Planel et al., 2006; Rozière, 2007; Rozière et al., 2009; El Hachem, 2010; El Hachem et al., 2012a, b). Tests are led at constant pH in controlled conditions. Hence, clarification about the dormant period (first stage) can be made when correlating mass/length measurements with kinetics of leached calcium/hydroxide ions. Tests are carried out at constant pH and temperature (e.g., 7.5 °C and 20 °C respectively). The pH is regulated by adding a nitric acid solution at 0.5 mol/L. The sodium sulfate solution is renewed every 15 mL of added acid. Hence, constant sulfate concentration is ensured during the test.

Mass and length are measured each time the solution is renewed. The monitoring of both ion concentration and mass/length can be carried out with the (same) mortar prisms. In this case, the prisms have gauge studs at both ends, in order to measure expansion with length change comparators, with an Invar cylinder as a reference length to zero the comparator. Before each mass and length measurement, the samples are wiped to remove the water from the sample surface. The samples remain saturated with dry surface. In parallel, the quantity of leached hydroxide ions is assessed through the added volumes of nitric acid solution and that of leached calcium by the titration of the solutions. The titration of the solution can also provide the sulfate ingress and this can be used to calibrate models. Cylindrical concrete specimens can also be used for the monitoring of the leached ions (see Fig. 4.35c (Rozière et al., 2009)).

Examples of results and setups used for tests combining mass/length change and ionic concentration measurements are presented in Fig. 4.35.

4.6.4 Experimental Study of Stresses Induced by in-Pore Crystallization: Fast Testing Method for Discriminating Cement Reactivity with Regard to ESA

A specific testing method, based on poromechanical approach, has been developed, aiming at better understanding of how in-pore crystallization induces (macroscopic) mechanical stresses at the scale of samples (*expansion stresses*) (Bui, 2016; Meulenyzer et al., 2020).

Compacted ground cement-based cylindrical samples are used where the material (cement paste) is well-hydrated. This model cement-based material is significantly more permeable than regular cement paste. Upon testing, the samples are placed in a so-called isochoric cell, which prevents the sample from expanding, but which is instrumented so that the mechanical stresses exerted by the sample (as a result of ettringite formation) on the walls of the cell can be measured. The test consists in flushing the sample in its isochoric cell with solutions of sodium sulfate

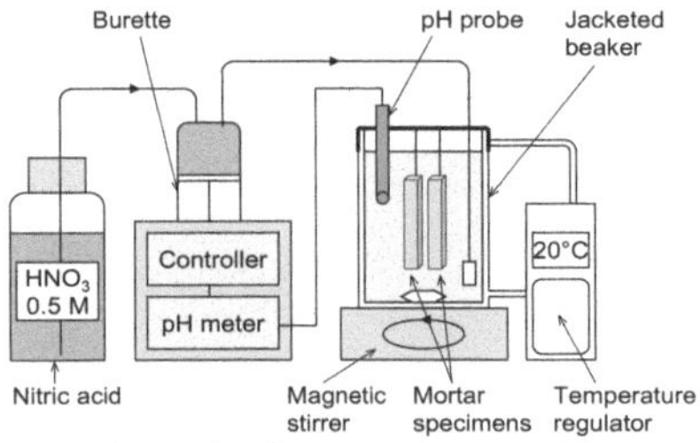

a) schematic diagram (mortar prisms) (El Hachem, 2010)

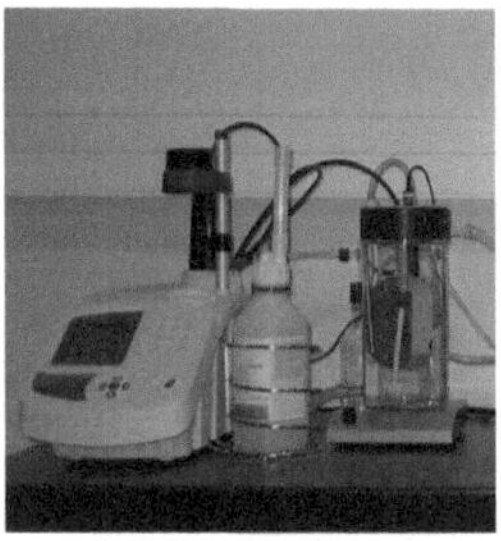

b) picture of setup (El Hachem, 2010)

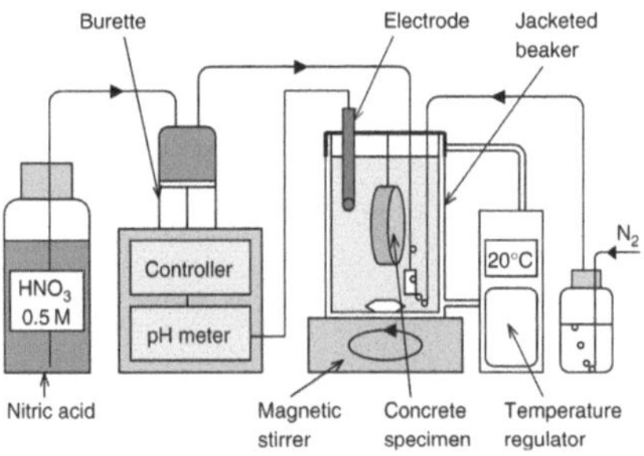

c) schematic diagram of leaching test on cylindrical concrete specimens (Rozière *et al.*, 2009)

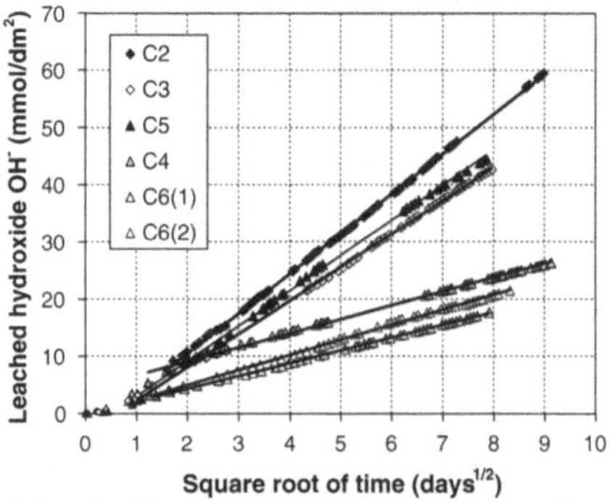

d) leached hydroxide ion amount measured on concrete specimens (Rozière *et al.*, 2009)

Fig. 4.35 Examples of results and setups used for tests combining mass/length change and ionic concentration measurements. *C3 (W/b = 0.50) contains 30%FA and C6 (W/b = 0.40) contains 62%GGBS, while C1 and C1B (W/b = 0.70), C2 (W/b = 0.50) and C4 (W/b = 0.40) are OPCs.* (**a**) schematic diagram (mortar prisms) (El Hachem, 2010) (**b**) picture of setup (El Hachem, 2010) (**c**) schematic diagram of leaching test on cylindrical concrete specimens (Rozière et al., 2009) (**d**) leached hydroxide ion amount measured on concrete specimens (Rozière et al., 2009)

of various concentrations and measuring the stress and the axial dimension evolution over time.

Bui tested a variety of mix-compositions (Bui, 2016). The stresses were revealed as independent of the concentration, but can be directly (linearly) correlated to the volume of ettringite crystals formed. In addition, the experimental results showed that the amount of sulfates retained in the sample was correlated to the duration of the injection and a duration of about 4 hours was suggested for the test procedure.

Such a testing strategy sounds promising, in order to discriminate (with the appropriate reproducibility) the sensitivity of binders (including those with SCMs) to ESA on the basis of a *very rapid lab test* (test duration of the order of a week) (Bui, 2016; Meulenyzer et al., 2020). An additional advantage of this test is to induce a more *homogeneous* response of the sample, and in particular a homogeneous sulfate distribution within the C-S-H gel microstructure. Moreover, additional parameters, such as permeability and sulfate binding capacity, can also be measured during the test (Meulenyzer et al., 2020).

4.6.5 Other Innovative Methods

Small cement paste pipes (i.e., hollow samples with wall thickness = 2.5 mm) have been used to assess the mechanical stresses causing the expansion (Ma et al., 2018). These samples can be submitted to longitudinal restraints before exposure to sulfate solutions. Free or restrained expansion, as well as the generated stresses, can be quantified.

A modified ring test for ESA testing can also be proposed to measure and assess longitudinal and volumetric deformations, the strength degradation and the sulfate diffusion coefficient by a single test on mortar samples and associated computations (Wan Kai Tai, 2016).

4.7 Final Remarks on the Review: Field Conditions Versus Accelerated Lab Tests

Final remarks on the review can be made, more specifically with regard to field conditions on the one hand, and to accelerated lab tests on the other hand, in order to focus on some significant aspects.

4.7.1 Field Conditions: Combined Actions

So far, durability design is based on (various) single processes. However, in the ESA process itself, [chemical reactions + transport processes (e.g., of moisture and ions)], which means reactive transport, takes place, along with mechanical consequences. In addition, in field conditions, ESA combined with other actions takes place. As a matter of fact, observations in field conditions (Menéndez et al., 2019; Menéndez et al., 2020b) reveal that in general a combination of actions (ESA and other processes) occurs, such as:

- concrete containing and/or exposed to sulfates at high temperature at early age (ESA + ISA/DEF),
- concrete exposed to sulfates, which contain alkalis and reactive siliceous aggregates (ESA + ASR),

- concrete exposed to both sulfate penetration and drying or to drying-wetting cycles (chemical form + physical form (induced by salt crystallization) of sulfate attack + hindered-shrinkage induced cracking in the near-surface zone),
- concrete in contact with low-pH water or groundwater (ESA + leaching),
- concrete in contact with seawater, where sodium sulfate concentration is 2.2% and pH $\approx$ 8, and where other ions are present, which yield multi-ionic transport and interactions (ESA + seawater attack),
- sulfate and chloride penetration in RC structures (ESA + chloride-induced reinforcement corrosion),
- sulfate and atmospheric carbon dioxide penetration in non-buried RC structures (ESA + carbonation-induced reinforcement corrosion),

and the mentioned actions can themselves be combined.

Hence, in accelerated lab tests, as well as in field conditions, concrete is submitted to attacks induced by at least the presence of sulfates and the low pH of the environmental medium (Saillio et al., 2016, 2019).

An enhanced degradation is observed in a great number of these combined actions, compared to the degradations induced by each single deteriorating process (e.g., ESA alone), as a result of synergetic effects (Bassuoni & Nehdi, 2009b).

For example, physical attack (salt scaling), which may take place in arid climates, is the most rapid and aggressive attack in structures, as already mentioned (see Sects. 4.1.1 and 4.2.3). In this case, concrete is exposed to sulfate penetration and to drying (or drying-wetting cycles). Drying of concrete may induce crystallization of salts and possible damage, according to the theory of the crystallization pressure (Scherer, 2004; Flatt & Scherer, 2008). In some situations, concrete is submitted to the chemical form (induced by ettringite/gypsum precipitation) + physical form (induced by salt crystallization) of sulfate attack at the same time, and an enhanced damage is observed (Menéndez et al., 2019). In addition, hindered-shrinkage induced cracking in the near-surface zone may enhance the damage.

With regard to the mechanism of the physical attack, wetting-drying cycles or cooling-heating cycles induce in particular repeated dissolution-recrystallization and phase transitions of sulfate salts (mainly between thenardite and mirabilite) at each cycle, which yield volume changes and thus fatigue in specimens/structures. Stresses sufficiently high to overcome the tensile strength of the material may be developed and the salts may exert a sufficient force to bring about disintegration of the material (Bonnell & Nottage, 1939) and to result in progressive spalling/scaling of the concrete near-surface zone (Scherer, 2004; Nehdi et al., 2014; Zhutovsky & Hooton, 2017).

Accelerated and enhanced degradation of the hydration products, as a result of wetting and drying cycles or of cyclic temperature changes, has been reported in the literature (see e.g., Bassuoni & Nehdi, 2009b; Gao et al., 2013; Jabbour, 2019), compared to the degradations induced by ESA alone. The buildup of salt crystallization pressure in the zones of structures which are not submerged may yield poor performance of concrete containing high amounts of GGBS (e.g., 85%) (Gruyaert et al., 2012). The most rapid and aggressive form of attack occurs in the case of wick action, where the rapid ingress of sodium and sulfate ions with the liquid phase induces crystal growth, followed by crystallization of salts in pores and (hindered) expansion due to phase changes (induced by temperature or RH fluctuations) (Zhutovsky & Hooton, 2017). Likewise, evaporation was confirmed as a key mechanism in the thaumasite-form of ESA (Mittermayr et al., 2013).

Moreover, the ESA effect on transport of (chloride) ions is now well known (see Sect. 4.2.2.2), and this is likely to promote reinforcement corrosion.

On the other hand, it is reported in the literature that the presence of chlorides in sulfate environments mitigates the ESA-induced deterioration (see Sect. 4.2.2.2). Likewise, in some conditions, (moderate) carbonation can induce a protective layer, likely to delay ESA.

The magnitude of these effects is difficult to correlate to the value of the actual parameters involved in ESA, and this may yield misleading explanations. For example, in the case of (ESA + ASR), cracks and enhanced pore volume induced by one degradation (e.g., by thaumasite-induced damage) promote penetration of agents enhancing the other (e.g., sodium with regard to ASR) (Menéndez et al., 2020a).

As a consequence:

- the analysis of field data is tricky, since coupled actions take place and both the environmental conditions and the material may evolve and change significantly in the course of time. In addition, the available information on the composition of the material and the history of the structure can be poor,
- it is difficult to separate the specific consequences of ESA from the (overall) degradation observed,
- the confusing origin of the degradation may yield irrelevant durability prediction or inappropriate/inefficient repair.

In front of these difficulties, **sensitivity analysis** can help to identify the major mechanism (depending on the experimental conditions).

4.7.2 Accelerated Lab Tests for Sulfate Resistance Testing: Current Situation

Most of experts/researchers in the ESA field agree that (existing) accelerated lab tests are **irrelevant** (e.g., with regard to the underlying mechanisms involved, to the oversimplified external solution used, etc.) with respect to field cases. However, they all use, develop or try to improve *accelerated* tests, for the reasons detailed in this report, and in particular to be able to assess ESA resistance within a reasonable time, since expansion kinetics are very slow in field conditions.

In addition and as a result, **a great number of different test methods exist or are under development, sometimes in a same country**. Applied test methods are however often variants of pre-existing and/or still-in-use standards (see Sects. 4.4,

4.5, and 4.6). In most of cases, the differences between the methods (and on the results) relate to:

- type of samples tested (e.g., on mortar vs. on concrete),
- curing and storing conditions of the samples,
- age of the samples,
- preconditioning procedure (if any),
- test protocol itself (e.g., full immersion vs. drying-immersion cycles),
- test duration: from a few weeks (e.g., 14 days (ASTM C452-15, 2015)) for the "very rapid tests" to several months in the general case (typically 6 months).

When designing/improving such accelerated lab tests, it is currently focused on:

- the most appropriate aggressive solution (e.g., comparison between solutions containing Na_2SO_4 and $MgSO_4$),
- the improvement of the reproducibility depending on the expansion level,
- the way of acceleration,
- the improvement of the representativeness with respect to field conditions,

even if the last two purposes can appear hardly compatible. However, one way to speed up the test, while improving its representativeness, could be pH control (case of full immersion) and more generally maintaining the boundary conditions (B.C.).

It is possible to roughly summarize the situation, and in particular the **weaknesses/drawbacks** of the available accelerated lab tests for sulfate resistance testing, as follows:

- no clear purpose of the test(s) used and/or developed,
- oversimplified approach (e.g., **based on a single (length change) measurement**), particularly irrelevant when softening or disintegration of the matrix takes place,
- significant issues in the existing experimental procedures,
- difficulties with preconditioning procedures,
- possible large differences between lab test conditions and field conditions,
- **ESA mechanisms strongly affected by the test parameters** (e.g., respective contributions of ettringite formation /gypsum precipitation/ion leaching),
- lack of relevant standardized rapid tests,
- variety of results published in the literature obtained by means of different protocols (difficult comparison between results),
- very few results from wide-range RRT,
- very few statistically valid precision data (as a consequence of the two last items) and hence unknown reproducibility,
- tricky interpretation of test results,
- the current evaluation methods do not seem sensitive enough to discriminate sulfate-resisting cements/concretes for every type of binders. More specifically, unknown applicability of the tests to new binders such as low-CO_2 binders,
- accelerated lab tests cannot directly be used to predict field performance and hence cannot be used alone to decide acceptability of cementitious materials.

Such weaknesses/drawbacks explain the inability to reach until now final conclusion in view of standardization of a method of assessment of the resistance to ESA.

Nevertheless, it is worth also focussing on some **positive aspects** of the current situation:

- significant efforts have been made for several decades, in order to improve the approach/protocol and promising methods have been developed, in particular with regard to the acceleration of the processes,
- approach/protocols have in particular been proposed, in order to characterize **combined actions** (e.g., sulfate penetration and moisture transport), which mimic some field conditions, by means of accelerated lab tests,
- the "suitability" of the test methods to some SCM-mixtures has been extensively investigated. For example, the effect of the substitution ratio of MK, limestone or GGBS has been studied,
- the impact of the "acceleration" of the processes is now better understood.

4.8 General Conclusion and Guidance/Recommendations for Relevant Sulfate Resistance Testing (in Lab)

4.8.1 Summary of the Background

One can find, in the available standards, general recommendations, as well as local recommendations, based on exposure classes, for the selection of cements resistant to ESA. In particular, EN 197-1: 2012 gives specifications of sulfate resisting cements. However, such an approach is far from sufficient to insure durability of concrete structures in field conditions, since other material, structural and/or environmental parameters may affect the overall resistance. The issue is enhanced when using non-traditional constituents or in the case of uncommon aggressive environments.

Faced to the need to:

- improve the available test procedures,
- develop suitable tools within the framework of performance-based approaches of durability of (reinforced) concrete structures,
- adapt the tools to new types of binders (e.g., low-CO_2 binders),

recommendations can be proposed, in order to continue the work towards the elaboration of a **relevant methodology for sulfate resistance testing in lab on both mortar specimens and concrete specimens**. In the near future, EN 206 could refer to such a methodology.

The guidance and recommendations proposed in this STAR by RILEM TC 251-SRT are mainly based on the:

- review of the literature available, in particular with regard to the most recent advances and new insights on the topic,
- worldwide experience, both in lab and in field, gathered until now,

which are detailed in the three chapters of this STAR.

4.8.2 Requirements for an Improved Testing Methodology

According to the literature, until now the **overall 1-D deformation of bars/prisms** is the parameter the most often measured during accelerated tests. In addition, this is the indicator of degradation the most widely used in the practice and adopted in existing standards (along with specifications on the expansion limit).

However, it has been in particular demonstrated for a long time (see, e.g., CEN/TR 15697, 2008; Kunther et al., 2013) that this approach is not sufficient for relevant assessment of ESA resistance and that its limitation is enhanced in the case of mixtures with SCMs. In this last case, samples can be (severely) damaged without measurable macroscopic deformation. The degradation occurs rather through surface mass loss (Chabrelie, 2010; Yu et al., 2013). In addition, significant surface degradation (e.g., induced by M-S-H formation), but lower expansion, can be observed in the case of $MgSO_4$ attack. Hence, such a poor resistance of the material to ESA cannot be captured by the single measurement of (1-D) deformations (Kunther et al., 2013).

In addition, one needs to keep in mind that ESA induces *heterogeneous* phase assemblage (e.g., so-called *zonation* in 1-D conditions) and hence differential expansion within the test specimen. However, only an *average* response of the specimen is recorded when measuring 1-D or even volumetric deformations. This fact limits the capability of results only based on this measurement to mimic the real phenomena.

The measurement of 1-D or volumetric deformations of specimens should be complemented by other investigations. More generally, the evaluation of the resistance to ESA cannot be based on a single test, such as expansion measurement in "accelerated conditions", as proposed unsuccessfully until now.

It is worth reminding here that damage at the structural level (e.g., cracking and spalling of concrete, loss of bond at the steel-concrete interface etc.) results from crystallization pressures within the nanoporous C-S-H phase and from material properties (e.g., transport properties), whose averaged values are measured in lab at the macroscale.

The size and geometry of the test specimens or of the structural elements significantly influence the mechanical effects and the degradations observed (and thus the lab test results) (Suma & Santhanam, 2020). In addition, as previously mentioned, gradients of properties are observed when ESA takes place. This illustrates that ESA is a *structural* problem. Consequently, it will not be possible to address such a structural problem only with material parameters. Characterization capable of providing quantitative information on the gradients, as well as **"structural" parameters,** are also needed.

The possibility to relate the (applied) test(s) to in-field behavior should be clearly indicated.

As a consequence, the following general conclusion can be drawn up, with regard to testing the resistance to external sulfate attack: **a new approach, based on a whole set of tests and/or measurements** is required, which can account for the coupled **chemical**, **physical** (e.g., transport) and **mechanical** components of the phenomenon, as well as for its **multi-scale** feature. This yields in particular the need to directly test **concrete** samples, at least in some cases.

Moreover, the specificities of the various types of **environment**, which determine the processes to occur, need to be, as far as possible, taken in consideration in the assessment of the ESA resistance of materials from lab tests.

A more **global** approach is needed, likely to (also) assess surface damage, including both expansive (scaling) and softening types of destruction. Nevertheless, prioritization of the possible attacks is also required (Baroghel-Bouny et al., 2004).

The proposed approach is detailed in the next section.

4.8.3 Proposed Approach for an Improved Testing Methodology

The test method(s) should be selected according to the **purpose** of ESA testing and intended "use" of the test, which could be for example:

1. qualification of **cement/binder** (e.g., low-CO_2 binder) with regard to ESA resistance,
2. optimization of **concrete mixtures**,
3. evaluation of the "potential" resistance/durability of **concrete structures/foundations** to ESA and ranking of concretes. In this case, the representativeness of the test(s) with regard to the actual **field conditions** is particularly significant. For example, depending on the field conditions, a test which involves *wetting-drying cycles* (such as the one described in the Swiss standard (SIA 262/1, 2019), for which long experience has already been acquired) or a test which (only) involves *full immersion*, will be selected.
4. understanding/modelling of the phenomena,
5. prediction of long-term behaviour of concrete structures.

It is worth mentioning that in the two last cases (*cases 4 and 5*), modelling is needed and constitutes the main tool. The results of lab test(s) will only be "used" at a given step of the assessment by the model.

In case 1, test(s) on **mortar specimens** can be appropriate, whereas in *cases 2 and 3*, test(s) on **samples** cast with the **concrete(s)** regarded should be preferred. *In case 4* and *possibly in case 1*, **cement paste specimens** can be needed (see Sect.

4.5.5. and (Ferraris et al., 2018), Sect. 4.6.4. and (Bui, 2016; Meulenyzer et al., 2020), as well as (Gu, 2018)).

In case 3, the possible mechanism(s), which take place in field conditions, need to be clearly identified. These mechanisms can be:

- *precipitation of* ***ettringite*** *with possible apparent swelling of the hardened cement paste*. In this case, the assessment of the resistance can be done at least by means of **length/volume** change measurement of mortar samples (along with **mass loss** measurement),
- *formation of* ***other products*** *than ettringite (such as gypsum, M-S-H, thaumasite, …) possibly without swelling/cracking but with other types of damage (e.g., desintegration)*. In this case, other types of measurements should be added, such as **surface degradation** observation, **aggregate loss** observation, and/or quantification of species/products by **TGA, XRD** or [**SEM + EDS mapping**], etc.,
- *occurrence of* ***leaching***. In this case, the resulting **zonation** will be studied. It can be observed on the degraded specimens, first visually (detection of the leached depth), or more precisely by **SEM,** and quantified (e.g., by means of [**SEM + EDS mapping**] and/or **TGA** measurement) with the appropriate precision (see Sect. 4.8.3). Besides, the monitoring of leaching, based on measurement of the **concentration of calcium and/or hydroxyl ions** of the bath solution as a function of time, can be useful (see Sect. 4.6.3),
- ***Na_2SO_4 salt*** *crystallisation (so-called physical attack)*. In this last case, tests involving partial immersion, wetting-drying cycles or cooling-heating cycles, with Na_2SO_4 solution, which do yield salt crystallisation and which are likely to mimic efflorescences and field degradation by physical attack, are required (see (Zhutovsky & Hooton, 2017; Dhole et al., 2019; Jabbour, 2019; Yoshida, 2019)).

After identification, prioritization of the mechanisms and selection of the relevant parameters along with their "weight" is needed, in order to carry out the most appropriate investigations and to select properly the test programme.

Hence, the proposed approach will yield **two situations**:

- *situation 1*: a general (set of) simple test(s) is sufficient (limited number of cases),
- *situation 2*: some additional tests and/or specific protocols are required, for example in the case of low-CO_2 binders where specific curing/preconditioning may be required, or in the cases where the **actual exposure conditions** during the life of the structural concrete needs to be taken into account. The definition of these exposure conditions can refer to so-called "durability classes" or "exposure resistance classes". With regard to the associated experiments, the following suggestions can be made (see Sect. 4.7.1):
  - *environment with sulfates, with constant B.C.* ➔ test in full immersion with constant pH, sulfate concentration and temperature, and with deformation/mass measurement, are appropriate,
  - *environment with sulfates and (strong) drying* or wetting-drying cycles, where physical sulfate attack may occur ➔ test involving partial immersion and/or

wetting and drying cycles, … (Dhole et al., 2019; Jabbour, 2019; Yoshida, 2019), will be preferred,
- *other coupled attacks* ➔ quantification of ionic (e.g., sulfate) concentration and solid (e.g., gypsum, ettringite or monosulfate) content profiles will be needed, as well as measurement of moisture distribution, porosity and transport properties.

The detailed recommended methodology is described in the next section.

4.8.4 Detailed Recommended Methodology

4.8.4.1 Recommended Investigations and Parameters Selected

As explained in the previous section, **deformation/expansion** measurement should be complemented in the proposed methodology by other characterizations, in order to allow one to draw a proper evaluation of the resistance of cement/concrete to ESA.

Compressive strength and **weight loss** for example are now often measured in addition to expansion. These parameters can thus easily be recommended, in order to assess the ESA resistance.

An updated proposal for a set of tests/measurements may also include, e.g., **visual degradation scheme** or a more quantitative global assessment of the degradation thanks to assessment of physical properties by NDT (e.g., EIS (Bragança et al., 2014), UPV, or resonant frequency (Zhutovsky & Hooton, 2017; Hossack et al., 2020)). Such an approach is suitable to the monitoring/evaluation of real structures. In addition, general *durability indicators* (DIs), such as **porosity** and **transport properties** (physical parameters), as well as **$Ca(OH)_2$ content** (chemical parameter), need also to be assessed at least before and after the accelerated test (see below and (Baroghel-Bouny et al., 2004)). As a matter of fact, the mineralogical transformations of the cement paste, as well as cracking and aggregate loss, induced by ESA yield significant evolutions of the physical properties of the concrete, such as an increase in porosity and diffusion coefficients.

Consequently, the following set of parameters can be proposed.

Overall Assessment

- **Visual observations** and **rating** based on a scale for degradation, in particular in the case of field exposure (Menéndez et al., 2019). Additional microscopic analyses (e.g., by SEM/EDS, XRD or XRF) may help to better interpret (locally) visual observations,
- **1-D deformation or (volume) expansion**. It is worth recalling here that, since heterogeneous expansion takes place, only an "average" response will be actually measured in this case,
- **Mass loss**. The measurement of this (very sensitive) parameter constitutes a simple and efficient way to monitor the ESA process. It can help one, for example, to identify the main mechanism: the formation of products (such as ettringite)

can be distinguished from the loss of matter (in the case of leaching). However, various phenomena may affect this parameter, and the data observed will be a combination of them. Hence, interpretation of the results may be tricky. However, it is worth mentioning here that in general a direct relationship is found between evolutions of expansion and mass loss (Menéndez et al., 2019).

Chemical Parameters

- **pH**, and if possible conductivity and composition of the pore solution. For example, the adsorption of sulfates can be monitored from the composition of the pore solution,
- **phase assemblage** (e.g., ettringite, thaumasite, gypsum, monosulfate, C-S-H, $Ca(OH)_2$). As a matter of fact, $Ca(OH)_2$ sounds as a major parameter in ESA mechanisms, since it is involved in carbonation, leaching ($Ca(OH)_2$ has been identified as the main leached mineral at least in lab conditions) and ettringite formation, which can all take place during ESA in lab or field conditions. Such data can be obtained by means of thermodynamic modelling (Lothenbach & Winnefeld, 2006; Lothenbach et al., 2010). In addition, during the "first" (non deleterious) phase of the attack, ettringite precipitation around aggregates, in big voids or cracks, without significant swelling, can be identified, e.g., by SEM coupled with EDS mapping. With regard to the "second" phase of the attack, methods which allow detection of sulfates in the C-S-H gel are needed. Quantitative EDS plots can then be obtained (see, e.g., Yu et al., 2012; Yu et al., 2013; Boudache et al., 2020). Such a detailed microstructural investigation will be carried out only when required,
- **ionic** (e.g., sulfate) **concentration** and **solid** (e.g., gypsum, ettringite, monosulfate) **content profiles**. They can be assessed, e.g., by means of XRD or TGA measurement (see Figs. 4.9 and 4.34d and (Santillán et al., 2020)). Profile measurements will allow one to quantify the *gradients*, which can appear during the test and to identify the "zonation" (including degraded zone, transition zone and non-affected core, see Fig. 4.36 (Gollop & Taylor, 1995; Whittaker & Black, 2015)). As a matter of fact, such a formation of zones has been reported many times in the literature (Gollop & Taylor, 1995) and has been recently confirmed by investigations on field test specimens (Chabrelie, 2010) and modeling (Idiart et al., 2011; Cefis & Comi, 2017). Thus, sulfur, SO_3 or sulfate distributions and

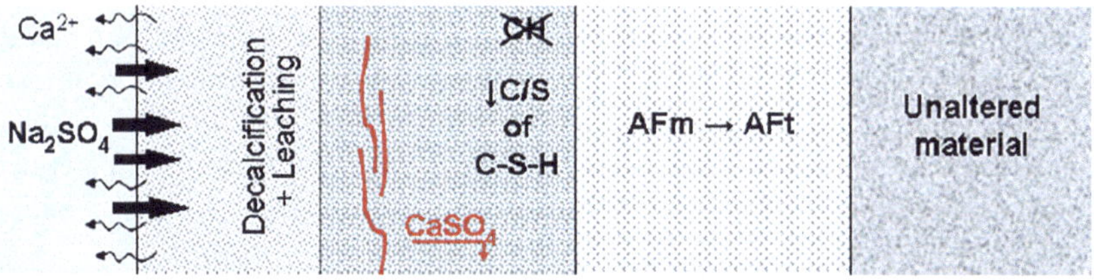

Fig. 4.36 Scheme of the various degraded zones of a Portland cement paste sample immersed in a Na_2SO_4 solution at T = 20 °C (case of ESA combined with leaching). (After Gollop & Taylor, 1995)

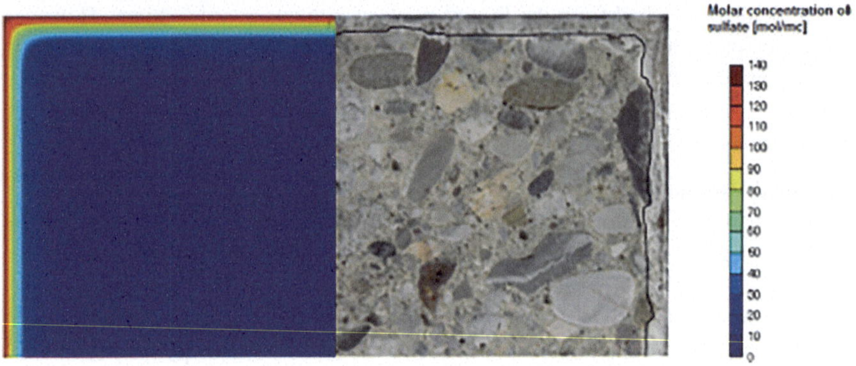

Fig. 4.37 Penetration front of sulfate after 400 days of immersion in a 10%-Na_2SO_4 solution: numerical simulation (left) and experimental profile (right). (After Cefis, 2016; Cefis & Comi, 2017)

the zonation can be observed on degraded specimens by SEM and quantified, e.g., by means of XRD, XRF, EDS or EPMA mapping, and/or TGA measurement, depending on the precision needed (see, e.g., Gollop & Taylor, 1995; Chabrelie, 2010; El Hachem et al., 2012a, b; Yu et al., 2013; Ma et al., 2018; Yoshida, 2019). The profile evolution as a function of time is likely to help one to quantify the progress of the degradations, in particular in mixtures where they cannot be monitored through expansion measurement (e.g., low-clinker binders),

- **sulfate binding capacity**. Such a quantification (e.g., through the measurement of binding isotherms) can be needed, e.g., in order to compare the binding capacity of various types of binders,
- outward (e.g., OH^- and Ca^{2+}) and inward (e.g., SO_4^{2-}) **ionic flows** can also be assessed, in order to quantify in particular leaching,
- **"front" of sulfate penetration or of degradation**. The progress of this front can be monitored by means of simple colorimetric methods based on drying (visual assessment of the penetration depth, in the case of 1-D geometry), as this is done, e.g., for chloride ingress in EN 12390-18+A1: 2024 or carbonation (see Fig. 4.37, (Cefis, 2016; Jabbour, 2019)). However, one has to keep in mind that drying will influence the location of the actual sulfate front, and such methods still need to be validated by means numerical simulations or other experimental methods.

Mechanical Parameters

- **Compressive strength**. Besides the usual standard (destructive) tests on cubes or cylinders, carried out in lab, and in view of field application, the ultrasonic pulse velocity (UPV) method can be used as a NDT method. This technique allows one to estimate the (decay) of the compressive strength of concrete (through the ratio of squared velocities in deteriorated and initial conditions (Zhutovsky & Hooton, 2017)),
- **Tensile strength**. The tensile or flexural strength can be preferred, since this parameter is more appropriate/sensitive than compressive strength to reflect the near-surface damage of concrete induced by ESA,

- **Dynamic modulus of elasticity**. Measurements of the dynamic E-modulus at different stages of the test (e.g., before, in the course and after exposure to sulfate solution) are likely to help one to link expansion and mechanical damage (quantified here by the decrease in this property) or to identify softening-type of damage. This can further help one to define relevant limit values for expansion (Loser & Leemann, 2016; Zhutovsky & Hooton, 2017) (see Sect. 4.9.3). The dynamic E-modulus can be assessed through the measurement of the resonant frequency (Zhutovsky & Hooton, 2017; Hossack et al., 2020),
- **Crack network**. Quantification of the (micro)crack network (density, opening, orientation, …) can be useful to characterize and quantify the (mechanical) damage in a direct manner. More precisely, the "take-off" phenomenon observed on expansion curves, would correspond to crack formation, according to (Chabrelie, 2010). Among the various techniques applicable, X-ray microtomography can be relevant (Ma et al., 2018; Boudache et al., 2020; Suma & Santhanam, 2020).

Physical Parameters

- ***Resistivity***. In the case of investigation of ESA coupled with rebar corrosion, electrical resistivity measurement would be particularly useful along with measurements and monitoring of more specific parameters to corrosion (requiring accessible bar) in RC test blocks or existing structures. Such a test has the additional advantages to be simple, quick and applicable in field conditions (see EN 12390-19:2023),
- ***Bulk water-accessible and/or MIP porosity***. The porosity is a global indication of the (residual) quality of the material. Hence, its assessment will be very useful. Since ESA may affect both the capillary and the C-S-H pore systems, and since deformations may be affected in turn by the void system, additional parameters such as the pore size distribution (PSD) will also be useful. PSD quantification can be useful in particular in order to investigate and distinguish between, e.g., filling of some voids/pores (induced by precipitation of reaction products) or coarsening of the void system (e.g., due to alkali and calcium leaching which may yield portlandite/C-S-H dissolution, delamination or cracking) (Ma et al., 2018; Gu, 2018; Gu et al., 2019; Jabbour et al., 2019),
- ***Transport properties***. Measurement of transport properties (e.g., chloride diffusion coefficient, as used in (Messad, 2009) or in (Jabbour et al., 2019), see also EN 12390-11:2015) seems quite relevant in the case of ESA (see Fig. 4.23b). As a matter of fact, the attack starts with ingress of sulfates and possibly outward movement of other ions (e.g., alkalis and OH^-) whose magnitude depends on the test conditions. In addition, chemical reactions can only involve locally available constituents. Moreover, the microstructural changes induced and previously mentioned will affect these properties. It is worth mentioning here that the ASTM C1202-19 test method (2019) has been proposed by authors within this framework, as a so-called rapid permeability test (Stephens & Carrasquillo, 2000),
- ***Degree of saturation*** *of the samples* (in the case of non-saturated conditions and when moisture transport takes place along with ESA). Mass measurement can be

used in particular for the assessment of an average value of the degree of saturation, while NDT can be preferred for the measurement of the moisture content distribution in the sample (Baroghel-Bouny et al., 2004).

4.8.4.2 General Requirements for Testing

The experiments need to be based as far as possible on standardized test procedures (when available) or at least on validated ones (for which some precision data are known). It is worth stressing that "sophisticated" measurements can be needed for the quantification of some of the additional parameters, with respect to deformations (in particular in cases 3 and 4). On the other hand, the cost of the test(s) is also a parameter to be accounted for.

Hence, once the types of test/measurement selected, the associated test method(s) should meet the following requirements:

- simple, robust, repeatable and reproducible,
- discriminating,
- with test duration acceptable for the practice (typically <3 months),
- carried out on mortar (or hcp) or concrete specimens, depending on the purpose/need of the test (see above),
- applicable to various types (or at least to the appropriate type) of binder(s) (e.g., blended cements such as CEM III, CEM V or incorporation of MK, limestone, etc.), and in particular to low-CO_2 binders. This means that the test methods include the suitable exposure and testing times, age …
- based on easy-to-build or "cheap" setup(s),
- with optimized size and geometry of samples with regard to the types of measurement needed,
- able to provide the appropriate *durability indicators* (specific to ESA resistance, see Sect. 4.9.3). These durability indicators refer to relevant chemical, physical and/or mechanical parameters, which are involved in the complex ESA mechanism(s), and which are needed for the application of, e.g., performance-based approaches of durability (including prediction models) and for which qualification criteria with respect to ESA resistance will be established.

4.8.4.3 Specific Requirements with Regard to Accelerated Lab Tests

It is worth reminding that when accelerated (performance) tests are needed for the assessment of the behaviour of concrete structures with regard to ESA, the mechanisms of degradation, as known according to the most recent advances, should be taken into account (e.g., swelling/cracking vs. softening/disintegration). In addition, the accelerated lab tests should be as much as possible "representative" of the behavior in the considered field conditions.

In particular, since these mechanisms, as well as the nature, relative proportions and chronology of occurrence of the reaction products (gypsum, ettringite and/or thaumasite) may differ according to the material and the (accelerated) test conditions (e.g., sulfate concentration, pH, size and geometry of the samples, …), the selected test method should involve the same mechanism, and respect the same **chronological occurrence** of the products formed, as the one that takes place in these field conditions. For example, **gypsum** (with respect to **ettringite**) formation is favoured when the sulfate concentration is increased. Likewise, **thaumasite** formation is favoured by low temperature ($T \leq 5$ °C) and the presence of carbonates (e.g., carbonate-rich aggregates).

In addition, the test method should involve the same **coupling** with other processes (such as leaching, penetration of chlorides and/or carbonation) as in the considered field conditions.

Before any "extrapolation" of accelerated lab test results, in view of contribution to predict the long-term behavior of actual structures, the relationship between the *applied test conditions* and the *precise exposure conditions* of the studied structure should have been properly established.

The following specific remarks can be added:

- *elevating the curing temperature* (before the accelerated test itself), in order to accelerate the hydration process, to reach as soon as possible a rather stable chemical state and to reach an appropriate degree of reaction, in particular in the case of slowly reacting binders. If this curing procedure is not applied, 90-day wet curing at $T = 20$ °C can be recommended,
- *acceleration* should be applied *without changing the mechanism(s)* and in particular the nature and relative proportions of the reaction products with respect to field conditions. In particular, it seems better to *avoid a too high sulfate concentration* in the exposure solution. Such high concentrations would induce marked differences in, e.g., the phase assemblage in the accelerated test conditions compared to field conditions, and differences in the crystallization pressures (and therefore in the interplay between chemical, physical and mechanical components of the process),
- *pre-drying* of the specimens (before full immersion) can also be used, in order to favor sulfate penetration (through capillary absorption),
- *preconditioning through [drying + saturation by sulfate solution under vacuum]* sounds also efficient to promote ESA (Messad, 2009; Aguayo et al., 2020; Boudache et al., 2020),
- *drying-wetting cycles* (more precisely drying-immersion cycles) can be used, in order to reduce the test length and favor gradient mitigation, as well as to better "represent" given field conditions. **An enhanced penetration and consecutive degradation** will then be observed. However, the analysis of the test results will be trickier,
- the pH is a key parameter in the ESA phenomenon. In particular, it is worth keeping in mind that maintaining a *low pH during the test will favour leaching.*

Therefore, the **pH value during the test**, as well as its eventual control, need to be select in accordance with the purpose of the test,

- with regard to the specimen configuration, **cylindrical specimens** (instead of prisms) should be preferred, in order to avoid edge effects,
- in the tests involving diffusion and/or capillarity, **1-D configuration** should be preferred to facilitate the analysis of the results and their use for the validation of (prediction) models. As a matter of fact, most of models are developed considering 1-D transport. 1-D tests can be carried out by sealing the non-exposed surfaces of the specimens (e.g., circumference of cylinder),
- the *test duration* will be between 6 and 12 months.

4.8.4.4 Examples of Recommended Experimental Conditions for Lab Tests Dedicated to Cement/Binder Evaluation and to Concrete Evaluation

Examples of recommended experimental conditions (storage conditions, concentration and volume of the contact solution, sample configuration, pre-curing, …), for accelerated lab tests dedicated to, on the one hand cement/binder evaluation, and on the other hand, concrete evaluation, are summarized in Tables 4.7 and 4.8, respectively.

The recommendations have been established in relation with the difficulties encountered with existing test methods. The selection of the options is in particular based on the conclusions of previous works carried out within the framework of:

- CEN/TC51-TC104/WG12/TG1 "*Sulfate resistance testing*" (e.g., round robin tests carried out on mortar specimens reported in the CEN/TR 15697 document published in 2008 (CEN/TR 15697, 2008)),
- NANOCEM European Consortium and associated ITN-MC projects dedicated to sulfate attack (Nanocem, 2004–2021),
- the discussions at the workshops hold by RILEM TC 251-SRT "*Sulfate Resistance Testing*",
- the literature review and investigations reported in the three chapters of this STAR.

Nevertheless, these proposed conditions need to be further investigated, validated and/or refined (e.g., through RRTs) within the European standardization framework.

Table 4.7 Examples of recommended experimental conditions for lab tests on mortar specimens dedicated to **cement/binder** evaluation

Issue	Difficulties with existing test procedures	Notes on candidate test methods	Selected options
Superficial phenomenon: There is evidence of inhomogeneity of the degraded test specimen The applied method should be sensitive to the (near-surface) degraded part of the specimen	Where occurs, expansion is due to the exposed surface only, not to the core of the specimen Thaumasite-based mechanism does not yield expansion but softening. Deformation measurement will be irrelevant in this case *Compressive strength* tests, as well as most of the other lab tests, are designed for homogeneous materials, without a core/shell macrostructure Visual inspection may be subjective	Tests sensitive to the surface: - *Flexural strength*. Most of the specimen section should be sound; only one surface should be degraded to a certain extent. In those conditions, the shape of the specimen will not be fundamentally affected while the poorer tensile strength of the degraded surface could make the flexural strength decreasing - *Mass loss*. Mass disaggregation may not occur (e.g., in case of thaumasite formation) Mass loss is influenced by operator handling and may be counteracted by mass gain	Determination of the *flexural strength* (as described in EN 196-1: 2016) One degraded surface only, to be tested in the extrados
Specimen shape and dimensions: Shape of the test specimen can strongly be affected by degradation. In those conditions, *mechanical tests* may be hardly applicable	Irregular shapes may be tested with unknown consequences on the reliability of the test results In the measurement of 1-D deformations, small specimens enhance expansion but may induce higher variability. However, they provide earlier and better detection Specimens with small sections will be deviating more from the nominal shape after degradation. This would reduce the applicability of the *mechanical test* and 1-D deformation measurements	*Mechanical tests* require tight control of the dimensions of the sample When using 40×40 mm^2 section, the test will not be rapid Squared sections prisms are attacked preferentially on the corners. However, corners are not influencing the *flexural strength* determination The flexural strength level is controlled by the defects on the extrados. When reducing the scale, the flexural strength is expected to be less sensitive in the detection of the defects induced by the attack	Preliminary check of the dimensions of each specimen and surface grinding where needed, in order to agree with requirements on tolerance, flatness and perpendicularity (see § "moulds" in EN 196-1: 2016) $40 \times 40 \times 160$ mm^3 (see § "preparation of test specimens" in EN 196-1: 2016)

(continued)

Table 4.7 (continued)

Issue	Difficulties with existing test procedures	Notes on candidate test methods	Selected options
Composition: Standardized composition is favoured	Technical disadvantages reported in (CEN/TR 15697, 2008) for paste/concrete tests High W/B, in order to accelerate the process and reduce the test duration, may depart from values specified in EN 206: 2013	Tests on standard mortar specimens are usually more rapid than on concrete samples	Recommended composition for OPC mortar: W/C = 0.50; sand/cement = 3 (see § "mortar constituents" and § "preparation of mortar" in EN 196-1: 2016)
Exposure conditions—Full immersion test: Most of the specimen section should be sound. In the simplest test, the exposure solution contains SO_4 anions only, and the concentration does not change during the test When testing relates to specific environment/case, the exposure solution (e.g., concentrations) could be close to the field conditions regarded. However, such an approach is more appropriate to concrete evaluation (see Table 4.8)	Without control, the pH will increase due to leaching and the SO_4 concentration of the exposure solution will decrease with time Most of the existing procedures use ion concentrations much higher than those classified for water by EN 206: 2013 Additional anions than SO_4^{2-} (e.g., chlorides as in seawater) will change the type of attack and the associated degradation mechanism(s). The presence of chlorides mitigates ESA The nature of the cation (e.g., calcium, sodium or magnesium) affects the mechanism(s) of degradation Using high concentrations of alkaline cations may cause ASR, with overlapping mechanisms	A full immersion test is more robust and reproducible (full immersion test is recommended in (CEN/TR 15697, 2008)) Controlling the pH within the range 7–8 accelerates the attack Increase in the frequency of the solution renewal, which helps maintaining large ionic concentration gradient, enhances the ESA process	Keep the pH constant (e.g., by adding H_2SO_4) Keep the SO_4^{2-} concentration constant in the solution Keep the volume of the exposure solution constant The use of intermediate sulfate concentration enhances the ESA process without significantly changing the mechanisms

Table 4.7 (continued)

Issue	Difficulties with existing test procedures	Notes on candidate test methods	Selected options
Exposure conditions—Partial immersion test or drying-immersion cycles: The possible additional heterogeneities should be reduced as far as possible Unnecessary deviations from the nominal shape should be avoided	Evaporation may affect the volume of the exposure solution Partial immersion increases the inhomogeneity of the degraded specimen Specimen borders are more degraded and usually make the specimens deviating more from the nominal shape The air-exposed surfaces are subjected to different/additional phenomena (e.g., drying and hence possibly salt crystallisation) than the submerged one The use of partial immersion or wetting-drying cycles induces additional variables to the problem, which need to be accurately quantified	Partial immersion or dying-immersion cycles will accelerate sulfate ingress, since penetration is mainly driven in this case by capillary suction. The ESA process will be enhanced Combined mechanisms will take place In the case of partial immersion or drying-immersion cycles, protection of surfaces can be applied, before exposure, by means of epoxy resins or equivalent methods, in order to ensure 1-D exposure and 1-D drying	Such exposure conditions will be selected in particular in order to reproduce physical sulfate attack Appropriate and well-defined liquid volume/surface (e.g., 20 L/m^2) Ensure 1-D exposure (and 1-D drying) Unprotected surface upwards
Temperature control: Different mechanisms occur at different temperatures **Thaumasite** formation is favoured at low temperature in the presence of carbonates (and hence by carbonation) The carbonate ion equilibrium is regulated by the pH		Not clear/systematic effect of high temperature. High temperature will enhance transport and sulfate component solubility and thus may accelerate the ESA process. However, ettringite is unstable from a given temperature and crystallisation pressures may be lower **Thaumasite** formation is favored at T < 5 °C and pH > 10.5	Two types of tests can be proposed: (a) T = 20–23 °C—pH = 8 ± 0.5 (which favours **ettringite** formation), (b) T = 5 °C—pH = 11 ± 0.5 (which favours **thaumasite** formation)

(continued)

Table 4.7 (continued)

Issue	Difficulties with existing test procedures	Notes on candidate test methods	Selected options
Curing and preconditioning: The required degree of reaction needs to be reached, in particular in the case of low-clinker binders	High temperature accelerates but may change the "hydration" mechanisms, more particularly in the first 14 hours Uncontrolled carbonation can affect the precision of the test results	Short wet curing accelerates the ESA degradation but may underestimate the performances of mature low-clinker binders The curing procedure should hinder carbonation	After demoulding at 1 day, 90-day curing in de-carbonated water at T = 20 °C, protected from carbonation, can be recommended
Testing times			At least after 1 year and on 6 specimens

Table 4.8 Examples of recommended experimental conditions for lab tests dedicated to **concrete** evaluation

Issue	Difficulties with existing test procedures	Notes on candidate test methods	Selected options
Specimen shape and dimensions: Shape should not be altered too much	Angular specimens are preferentially degraded by ESA Evaporation, which occurs in the case of partial immersion or during wetting-drying cycles, may cause salt crystallization and consequently rapid degradation of the specimen Both phenomena have negative impact on the shape preservation	If the cylindrical surface is protected, the evaporation is limited to the upper base of the cylinder (1-D geometry) 1-D geometry allows one to assess visually the (evolution of the) *degraded depth* (Cefis, 2016; Jabbour, 2019) More detailed microstructural (additional) investigations will be carried, if needed	*Cylinder*, Ø150 mm × 300 mm Circumference surface protected Cylinders are also recommended for the measurement of durability-related properties This shape has the advantage of removing corners and thus of reducing the variability Cut a cross section through the cylinder, in order to assess the *degraded depth* The quality of the exposed surface should be improved by cutting the outer layer or coring the specimens

Table 4.8 (continued)

Issue	Difficulties with existing test procedures	Notes on candidate test methods	Selected options
Composition and detection method: Need to cover wide ranges of concrete qualities, even including low W/B		Different mechanisms can occur for different concrete compositions Various parts of the specimen can also be subjected to different degradation mechanisms *Tensile splitting*: Tensile stress occurs in the central part of the specimens; marginally affected by the surface degradation	Tensile splitting strength according to EN 12390-6: 2023
Exposure conditions: Higher need to accelerate the degradation process, when long storage of concrete specimens is not easy	Full immersion could require long exposure time before observable degradation	*Electrical pulse* may be applied, in order to enhance sulfate ingress and significantly accelerate the process	In order to accelerate the process - drying-immersion cycles or - partial immersion, can be used In this last case, ¼ cylinder height can be submerged. The (bottom) submerged base has to lay on supports, which make easier the flux through the exposed surface
Exposure solution: Specific cations need to be considered individually	Various types of exposure solutions can be used with various types and concentrations of cations Synergic effects can take place	The pH in the central part of the specimen will be marginally affected by the pH of the exposure solution The use of multiionic solutions such as seawater involves complex interactions with the cementitious matrix and hence yields difficulties in the interpretation of the test result	e.g., **6** or **34 g/L** $\mathbf{SO_4^{2-}}$ obtained with: *(1)* $\boldsymbol{Na_2SO_4}$, *(2)* $MgSO_4$, *(3)* $(NH4)_2SO_4$ depending on the purpose of the test

(continued)

Table 4.8 (continued)

Issue	Difficulties with existing test procedures	Notes on candidate test methods	Selected options
Curing and preconditioning: The required degree of reaction needs to be reached, in particular in the case of low-clinker binders		Curing at moderate temperature may impact the distribution of aluminate-containing phases, the crystals development and at the end the pore structure	90-day water curing can be recommended Samples can be preconditioned, e.g., through [drying + saturation by sulfate solution under vacuum], in order to "accelerate" the process
Testing times		The test duration is usually between 6 and 12 months	e.g., after 1-, 3-, 6- and 12-month exposure on one specimen each one

4.9 Questions Still Open, Perspectives and Further Steps Needed: Adaptation of Tests/Models to Low-CO_2 Binders

4.9.1 Standardization and Round Robin Tests

The test method(s) proposed by RILEM TC 251-SRT need to be standardized in the near future, in order to ensure at least harmonized application of test procedure(s) in a great number of countries. For some types of tests, the first step will be to define a "universal" procedure, based on the existing variants. **Future standardization of the test method(s)** will be made within the framework of the re-formed CEN/TC51-TC104/WG12/TG1 "*Sulfate resistance to external sulfate attack*".

Some (repeatability) precision data associated with already published test methods can be found in the literature (see, e.g., Loser & Leemann, 2016; Ferraris et al., 2018). However, **inter-laboratory tests (RRT) associated with the proposed test method(s)**, involving a broad range of materials including low-CO_2 binders, are needed, in particular in order to get more relevant **repeatability/reproducibility data**, to be included in the future standard test method(s). For example, proper reproducibility, such as $\leq 15\%$, is required in usual standard test method(s) (see Sect. 4.1.2).

In addition, such RRTs will allow one to verify the robustness of less common test procedures, such as those based on sulfate ingress by capillary absorption, given their practical interest.

Moreover, (accelerating) preconditioning procedures, such as pre-drying, [drying + saturation with sulfate solution under vacuum] or drying-wetting cycles, need to be further investigated, in order to assess their robustness and their effects.

At the end, the RRT results will provide matter to refine the proposed approach and procedures and will provide matter to establish proper acceptance criteria.

No RRT has been carried out within the lifetime of RILEM TC 251-SRT. The suggested RRTs could hence be carried out within the framework of the re-formed CEN/TC51-TC104/WG12/TG1 "*Sulfate resistance to external sulfate attack*".

4.9.2 Prediction of Long-Term Behavior: Development and Improvement of Models

The approach proposed in this report based in particular on (accelerated lab) tests is not intended for *prediction of the structural behavior in actual field conditions*. Models will be needed, in particular in order to:

- help **understanding/confirming** mechanisms and their time of occurrence (by testing various scenarii/assumptions, ...),
- carry out **sensitivity analysis**, in order to identify the major parameters/ mechanisms,
- **predict** long-term behaviour (e.g., from short-term results and materials properties such as *durability indicators* (DIs), see Sects. 4.8 and 4.9.3). In this last case, (a) model(s) should be **proposed along with** the previously mentioned lab-test method(s).

More generally, it is worth noting that a combination of experiments and models will allow one to get a more reliable and rapid answer about long-term behavior and potential damage.

Nevertheless, few models are currently easily available. Recommendations can thus be made also with regard to the use, development and/or improvement of **models**:

- A multi-scale approach is needed,
- Coupled models are needed, since ESA involves **coupled chemical, physical (e.g., transport) and mechanical components** and since ESA occurs along with other processes in field conditions. Various types of models can be selected, such as:
  - **reactive transport models**. These advanced, accurate and relevant approaches have been proposed in the case of ESA, as for other degradation processes, in order to handle the complexity of the phenomena, which involve multiionic outward/inwards diffusion processes along with formation of ettringite, gypsum, etc. (see, e.g., HYTEC (van der Lee et al., 2003; Guillon & Moranville, 2004; Soive & Tran, 2017) or (Meulenyzer et al., 2020)). In this case, thermodynamic modeling coupled with kinetic equations, which describe the dissolution of the clinker, and databases are often used, in order to properly compute the evolution of the phase assemblage as a function of time and its effects on the pore structure (see, e.g., (Barbarulo, 2002) or GEMS (Lothenbach & Winnefeld, 2006)). In particular, the models need to account for the reaction kinetics, since they significantly affect the sulfate concentration profiles obtained after exposure,

- **chemo-mechanical models** (involving weak coupling with transport processes in fully or partially saturated conditions). The approach proposed by *Cefis* (see Fig. 4.37 and (Cefis, 2016; Cefis & Comi, 2017)) or the model developed at the meso-scale by *Idiart* (Idiart, 2009; Idiart et al., 2011) can be mentioned here. This last model is in particular able to predict the impact of the C_3A content on the degradation (cracking) and includes fracture-based constitutive laws. Description of expansion stress states, actual microcrack formation and decrease in elastic properties can be provided. Note that meso-scale models can bridge the material phenomena which take place at the micro-scale and the structural behaviour at the macro-scale. Nevertheless, oversimplified description of the chemical reactions (e.g., only based on reactions with aluminates) is often adopted,
- **poromechanical models.** This type of approach has been proposed, in order to simulate expansion of cementitious-material elements submitted to ESA (and also to DEF) (Flatt & Scherer, 2008; Gu, 2018). Such models are based on the surface-controlled crystal growth theory and physical-chemical properties, and can be coupled with any mechanical approach,

- In addition, literature review of the available models reveals **weaknesses in the mechanical aspects**, since coupling with the mechanical part is the trickiest and hence **fully coupled models** are still missing. Efforts should thus be made in the future on this aspect. Besides, better understanding of how ESA generates expansion and expansion stress is still required,
- The porosity and transport properties, initially included as constant input data, should be updated in the course of the computation, in order to account for pore structure and cracking evolution as a function of time,
- Extension of 1D to 2D or 3D modeling can also be useful, given the volumetric feature of the deformation and the heterogeneous feature of the damage observed. In addition, appropriate mesh needs to be selected.

Moreover, benchmark of models will be needed, including field conditions, in order to compare, validate and *in fine* select the most relevant tools, depending on the purpose.

4.9.3 Durability Indicators/Indexes (DIs) Specific to ESA

Specific *durability indicators/indexes* (DIs) to ESA remain to be defined, as previously done, e.g., for ASR (Baroghel-Bouny et al., 2004). Such DIs have to meet the following requirements:

- chemical, physical and/or mechanical parameters, which are relevant, i.e., which are involved in the complex ESA mechanisms,
- parameters involved in the basic equations and thus required as input data for (predictive) models,

- scientifically based, but also easily accessible parameters.

For example, it could be investigated if among the chemical parameters, the content of (hydrated) aluminates of the binder is an appropriate DI candidate, depending on the type of ESA. However, some information is still missing with regard to the material parameters involved in the ESA, in order to be able to propose relevant DIs specific to ESA. For example, the literature is scarce and contradictory on the role of C_4AF. Further research would hence be useful on this aspect.

Some test procedures (see Sects. 4.3 and 4.4) include a criterion on the expansion at (a) given time(s), in order to classify the cement as "sulfate resisting cement". However, experience still needs to be acquired on a broad range of (concrete) mixtures, before the establishment of relevant criteria/specifications associated with the mentioned DIs.

4.9.4 Better Understanding of the Mechanisms Involved in the Case of Low-CO_2 Binders

There is an increasing use of binary and even ternary blends in actual concrete mixtures, and more generally of various SCMs, by-products, recycled aggregates (Santillán et al., 2020), and/or local materials, in order to obtain **low-CO_2 binders**. These alternate solutions to CEM-I binders are needed. However such an evolution of the mix-design could increase, for the resulting materials, the risk of sensitivity to ESA. A lot of information is already available on SCM-materials (e.g., FA-materials), but investigations are needed on more innovative low-CO_2 mixtures.

In addition, a better understanding of the mechanisms involved in ESA in the case of low-CO_2 binders, or at least of the specificities of these materials, is required. For example, better understanding of GGBS-mixture behavior is needed, in order to explain in particular (see Sect. 4.2.4):

- the large deterioration observed after drying-immersion cycles, while good behaviour is generally observed after full-immersion lab tests,
- the damage in GGBS-concrete "field" samples, which are revealed as more eroded than CEM-I concrete samples (Chabrelie, 2010),
- the influence of the replacement ratio on the ESA resistance,
- the behaviour in the case of early-age exposure,
- and to better predict the effect of SCMs on the resistance to ESA.

4.9.5 In-Field Behavior: Development of Exposure Sites Dedicated to ESA

Exposure sites designed for the study of ESA in field conditions are scarce over the world (see, e.g., Figure 4.38 (Hossack et al., 2020)) and very few experimental data from long-term field exposure are available.

Trials, where concrete test specimens are exposed to "natural" conditions, are hence needed, in view of collecting more data and in order to:

- improve the knowledge of in-field behaviour,
- contribute to establish proper lab/field relationships,
- validate prediction models (see Sect. 4.9.2) in field conditions.

Fig. 4.38 Underground storage in field conditions (4–13 °C) in Canada (Hossack et al., 2020)

References

ACI 318-19. (2019). *Building code requirements for structural concrete and commentary*. ACI.

Aguayo, F., Funez, O. J., Drimalas, T., Folliard, K. J., & Lute, R. D. (2020). An alternative method to evaluate the sulphate resistance of cementitious binders. In E. Menéndez & V. Baroghel-Bouny (Eds.), *Proceedings of the workshop on external sulfate attack. Field aspects and lab tests (TESA 2018), 24–25 May 2018, Madrid, Spain* (RILEM Bookseries, Vol. 21, pp. 93–105). RILEM Publications.

Aköz, F., Türker, F., Koral, S., & Yüzer, N. (1999). Effects of raised temperature of sulfate solutions on the sulfate resistance of mortars with and without silica fume. *Cement and Concrete Research, 29*, 537–544.

Al-Amoudi, O. S. B., Maslehuddin, M., & Abdul-Al, Y. (1995). Role of chloride ions on expansion and strength reduction in plain and blended cement in sulphate environments. *Construction and Building Materials, 9*(1), 25–33.

Aleksic, M., & Hooton, R. D. (2011). *Development and standardization of the NIST rapid sulfate resistance test* (R&D Serial No. 3162). PCA Publication.

Alexander, M. G., Ballim, Y., & Stanish, K. (2008). A framework for use of durability indexes in performance-based design and specifications for reinforced concrete structures. *Materials and Structures, 41*, 921–936.

ASTM C1012/C1012M-18. (2018). *Standard test method for length change of hydraulic-cement mortars exposed to a sulfate solution*. ASTM International.

ASTM C1202-19. (2019). *Standard test method for electrical indication of concrete's ability to resist chloride ion penetration*. ASTM International.

ASTM C150/C150M-18. (2018). *Standard specification for Portland cement*. ASTM International.

ASTM C307-18. (2018). *Standard test method for tensile strength of chemical-resistant mortar, grouts, and monolithic surfacings*. ASTM International.

ASTM C452-15. (2015). *Standard test method for potential expansion of Portland-cement mortars exposed to sulfate*. ASTM International.

Barbarulo, R. (2002). *Comportement des matériaux cimentaires: Actions des sulfates et de la température [Behavior of cementitious materials: Actions of sulfates and temperature]* (in French). Ph.D. thesis, ENS Cachan, France.

Barbarulo, R., Peycelon, H., & Leclercq, S. (2007). Chemical equilibria between C–S–H and ettringite, at 20 and 85 °C. *Cement and Concrete Research, 37*, 1176–1181.

Barker, A. P., & Hobbs, D. W. (1999). Performance of Portland limestone cements in mortar prisms immersed in sulfate solutions at 5 °C. *Cement and Concrete Composites, 21*, 129–137.

Baroghel-Bouny, V., et al. (2004). *Concrete design for a given structure service life – Durability management with regard to reinforcement corrosion and alkali-silica reaction. State-of-the-art and guide for the implementation of a predictive performance approach based upon durability indicators* (AFGC Scientific and Technical Documents). AFGC (issue in French: 2004, issue in English: 2007), 240p.

Bassuoni, M. T., & Nehdi, M. L. (2009a). Durability of self-consolidating concrete to different exposure regimes of sodium sulfate attack. *Materials and Structures, 42*, 1039–1057.

Bassuoni, M. T., & Nehdi, M. L. (2009b). Durability of self-consolidating concrete to sulfate attack under combined cyclic environments and flexural loading. *Cement and Concrete Research, 39*(3), 206–226.

Bellmann, F., & Stark, J. (2007). Prevention of thaumasite formation in concrete exposed to sulphate attack. *Cement and Concrete Research, 37*, 1215–1222.

Bellmann, F., Möser, B., & Stark, J. (2006). Influence of sulfate solution concentration on the formation of gypsum in sulfate resistance test specimen. *Cement and Concrete Research, 36*(2), 358–363.

Biczok, I. (1967). *Concrete corrosion concrete protection*. Chemical Publishing.

Blanco-Varela, M., Aguilera, J., & Martinez-Ramirez, S. (2006). Effect of cement C_3A content, temperature and storage medium on thaumasite formation in carbonated mortars. *Cement and Concrete Research, 36*(4), 707–715.

Bonnell, D. G. R., & Nottage, M. E. (1939). Studies in porous materials with special reference to building materials: I. The crystallisation of salts in porous materials. *Journal of the Society of Chemical Industry, 58*, 16–21.

Boudache, S., Rozière, E., Loukili, A., & Colina, H. (2020). Influence of thermal preconditioning on the mechanism of external sulphate attack. In E. Menéndez & V. Baroghel-Bouny (Eds.), *Proceedings of the workshop on external sulfate attack. Field aspects and lab tests (TESA 2018), 24–25 May 2018, Madrid, Spain* (RILEM Bookseries, Vol. 21, pp. 71–80). RILEM Publications.

Bragança, M. O. G. P., Portella, K. F., Bonato, M. M., & Marino, C. E. B. (2014). Electrochemical impedance behavior of mortar subjected to a sulfate environment – A comparison with chloride exposure models. *Construction and Building Materials, 68*, 650–658.

Brown, P. W. (1981). An evaluation of sulfate resistance of cements in a controlled environment. *Cement and Concrete Research, 11*, 719–727.

Bui, N. N. (2016). *Expansion and stresses induced by crystallization in cement-based materials in presence of sulfates*. Ph.D. thesis, University of Paris-Est.

Cefis, N. (2016). *Mechanical effects of sulfate attack on concrete: Experimental characterization and modeling*. Ph.D. thesis, Politecnico di Milano, Italy.

Cefis, N., & Comi, C. (2017). Chemo-mechanical modelling of the external sulfate attack in concrete. *Cement and Concrete Research, 93*, 57–70.

CEN/TR 15697. (2008, April). *Cement – Performance testing for sulfate resistance – State of the art report*. CEN, 27p.

CEN/TR 16563. (2013, August). *Principles of the equivalent durability procedure*. CEN.

Chabrelie, A. (2010). *Mechanisms of degradation of concrete by external sulfate ions under laboratory and field conditions*. Ph.D. thesis, EPFL, Switzerland.

Cohen, M. D., & Mather, B. (1991). Sulfate attack on concrete – Research needs. *ACI Materials Journal, 88-M9*, 62–69.

Crammond, N. J. (2002). The occurrence of thaumasite in modern construction – A review. *Cement and Concrete Composites, 24*(3–4), 393–402.

Crammond, N. J., & Halliwell, M. A. (1995). *The thaumasite form of sulfate attack in concretes containing a source of carbonate ions – A microstructural overview* (ACI Symposium Publication, SP-154, pp. 357–380). ACI.

CSA A3004-08. (2008). Test methods and standard practices for cementitious materials for use in concrete and masonry. In *Cementitious materials compendium*. Canadian Standards Association.

De Weerdt, K., Lothenbach, B., & Geiker, M. R. (2019). Comparing chloride ingress from seawater and NaCl solution in Portland cement mortar. *Cement and Concrete Research, 115*, 80–89.

Dehwah, H. A. F. (2007). Effect of sulfate concentration and associated cation type on concrete deterioration and morphological changes in cement hydrates. *Construction and Building Materials, 21*, 29–39.

Dhole, R., Thomas, M. D. A., Folliard, K. J., & Drimalas, T. (2019). Chemical and physical sulfate attack on fly ash concrete mixture. *Materials Journal, 116*(4), 31–42.

DIBt. (n.d.). *German SVA test method – Test plan for the approval test of a cement as SR cement*. Deutsches Institut für Bautechnik (DIBt), Sachverständigenausschuss (SVA).

Divet, L., & Randriambololona, R. (1998). Delayed ettringite formation: the effect of temperature and basicity on the interaction of sulphate and C–S–H phase. *Cement and Concrete Research, 28*(3), 357–363.

Duval, R., & Hornain, H. (1992). Chapter 9: Concrete durability exposed to aggressive water. In J. Baron & J.-P. Ollivier (Eds.), *Durability of concretes (in French)*. Presses de l'ENPC, ENPC.

El Hachem, R. (2010). *Etude multi-critères de la dégradation des matériaux cimentaires par l'attaque sulfatique externe* [*Multi-criteria study of the degradation of cementitious materials by external sulfate attack*] (in French). Ph.D. thesis, Ecole Centrale de Nantes, France.

El Hachem, R., Rozière, E., Grondin, F., & Loukili, A. (2012a). New procedure to investigate external sulphate attack on cementitious materials. *Cement and Concrete Composites, 34*(3), 357–364.

El Hachem, R., Rozière, E., Grondin, F., & Loukili, A. (2012b). Multi-criteria analysis of the mechanism of degradation of Portland cement based mortars exposed to external sulphate attack. *Cement and Concrete Research, 42*(10), 1327–1335.

EN 12390-11. (2015, August). *Testing hardened concrete – Part 11: Determination of the chloride resistance of concrete, unidirectional diffusion*. CEN.

EN 12390-18+A1. (2024, October). *Testing Hardened Concrete – Part 18: Determination of the Chloride Migration Coefficient*. CEN.

EN 12390-19. (2023, May). *Testing hardened concrete – Part 19: Determination of electrical resistivity*. CEN.

EN 12390-3. (2019, June). *Testing hardened concrete – Part 3: Compressive strength of test specimens*. CEN.

EN 12390-6. (2023, November). *Testing hardened concrete – Part 6: Tensile splitting strength of test specimens*. CEN.

EN 196-1. (2016, April). *Methods of testing cements – Part 1: Determination of strength*. CEN.

EN 196-2. (2013, June). *Method of testing cement – Part 2: Chemical analysis of cement*. CEN.

EN 197-1. (2012, April). *Cement – Part 1: Composition, specifications and conformity criteria for common cements*. CEN.

EN 1992-2. (2005, October). *Eurocode 2 – Design of concrete structures – Concrete bridges – Design and detailing rules*. CEN.

EN 206. (2013, December). *Concrete – Specification, performance, production and conformity*. CEN.

Feldman, R. F., Beaudoin, J. J., & Philipose, K. E. (1991). Effect of cement blends on chloride and sulphate ion diffusion in concrete. *Il Cemento, 88*, 1–18.

Feng, P., Garboczi, E. J., Miao, C., & Bullard, J. W. (2015). Microstructural origins of cement paste degradation by external sulfate attack. *Construction and Building Materials, 96*, 391–403.

Fernàndez-Altable, V., Yu, C., Le Saoût, G., & Scrivener, K. (2011, July 3–8). Characterisation of slag blends and correlation with their sulphate resistance in static and semi-dynamic conditions. In *Proceedings of the XIII international congress on the chemistry of cement*, Madrid, Spain.

Ferraris, C. F., Clifton, J. R., Stutzman, P. E., & Garboczi, E. J. (1997). Mechanisms of degradation of Portland cement-based systems by sulfate attack. In K. L. Scrivener & J. F. Young (Eds.), *Mechanisms of chemical degradation of cement-based systems* (pp. 185–192). E & FN Spon.

Ferraris, C. F., Stutzman, P. E., Peltz, M., & Winpigler, J. (2005). Developing a more rapid test to assess sulfate resistance of hydraulic cements. *Journal of Research of the National Institute of Standards and Technology, 110*(5), 529–540.

Ferraris, C. F., Stutzman, P. E., & Snyder, K. (2006). *Sulfate resistance of concrete: A new approach* (R&D Serial No. 2486). PCA Publication.

Ferraris, C. F., Stutzman, P. E., & Peltz, M. (2018). Performance testing of hydraulic cements: Measuring sulfate resistance. *Journal of Research of the National Institute of Standards and Technology, 123*, 123010.

Flatt, R. J., & Scherer, G. W. (2008). Thermodynamics of crystallization stresses of ettringite crystals. *Cement and Concrete Research, 38*, 325–336.

François, R., Francy, O., Caré, S., Baroghel-Bouny, V., Lovera, P., & Richet, C. (2001). Mesure du coefficient de diffusion des chlorures, Comparaison entre régime permanent et régime transitoire [Measurement of the chloride diffusion coefficient, comparison between steady-state and non-steady-state regime] (in French). *Revue Française De Génie Civil, 5*(2–3), 309–329.

Fu, Y., Xie, P., Gu, P., & Beaudoin, J. J. (1994). Effect of temperature on sulphate adsorption/desorption by tricalcium silicate hydrates. *Cement and Concrete Research, 24*(8), 1428–1432.

Gao, J., Yu, Z., Song, L., Wang, T., & Wei, S. (2013). Durability of concrete exposed to sulfate attack under flexural loading and drying-wetting cycles. *Construction and Building Materials, 39*, 33–38.

GB/T 50082. (2009). *Standard for test methods of long-term performance and durability of ordinary concrete*. National Standard, China Building Materials Federation.

GB/T 749. (2008). *Test method for determining capability of resisting sulfate corrosion of cement*. National Standard, China Building Materials Federation.

Glasser, F. P. (2009). The thermodynamics of attack on Portland cement with special reference to sulfate. In M. G. Alexander & A. Bertron (Eds.), *Proceedings of the RILEM TC 211-PAE final conference on concrete in aggressive aqueous environments, performance, testing and modeling, June 3–5, 2009, Toulouse, France* (PRO 63, Vol. 1, pp. 3–17). RILEM Publications.

Gollop, R. S., & Taylor, H. F. W. (1995). Microstructural and microanalytical studies of sulfate attack. III. Sulfate resisting Portland cement: Reactions with sodium and magnesium sulfate solutions. *Cement and Concrete Research, 25*(7), 1581–1590.

Gruyaert, E., van den Heede, P., Maes, M., & De Belie, N. (2012). Investigation of the influence of blast-furnace slag on the resistance of concrete against organic acid or sulphate attack by means of accelerated degradation tests. *Cement and Concrete Research, 42*, 173–185.

Gu, Y. (2018, December). *Experimental pore scale analysis and mechanical modeling of cement-based materials submitted to delayed ettringite formation and external sulfate attacks*. Ph.D. thesis, University of Paris-Est, France, 295p.

Gu, Y., Martin, R.-P., Omikrine-Metalssi, O., Fen-Chong, T., & Dangla, P. (2019). Pore size analyses of cement paste exposed to external sulfate attack and delayed ettringite formation. *Cement and Concrete Research, 120*, 105766.

Guillon, E. (2004). *Durabilité des matériaux cimentaires – Modélisation de l'influence des équilibres physico-chimiques sur la microstructure et les propriétés mécaniques résiduelles* [*Durability of cementitious materials – Modeling the influence of physico-chemical equilibria on microstructure and residual mechanical properties*] (in French). Ph.D. thesis, Ecole Normale Supérieure de Cachan, France.

Guillon, E., & Moranville, M. (2004, March, 22–24). Physical and chemical modeling of Portland cement pastes under seawater attack. In *Proceedings of the ACBM/RILEM Symposium "Advances in concrete through science and engineering"*, Evanston, IL, USA.

Hartmann, P., Gilbert, M.-J., & Traber, D. (2013, December). *Hydrogeologische beobachtungen in den Bözbergtunnels* [*Hydrogeological observations in the Bözberg tunnels*] (in German), Arbeitsbericht NAB 13–15. http://www.nagra.ch

Haufe, J., & Vollpracht, A. (2019). Tensile strength of concrete exposed to sulfate attack. *Cement and Concrete Research, 116*, 81–88.

Henrique de Bem, D., & Medeiros, R. A., Jr. (2020). Accelerated tests to measure efflorescence and effect of Pozzolans. *ACI Materials Journal, 117*(5), 113–124.

Hernández, M. S., Guerrero, A., Goñi, S., & Lorenzo, M. P. (1997). Effect of the temperature on the leaching performance of the cement-based immobilization systems: Sulfate and chloride. *Cement and Concrete Research, 27*(4), 515–524.

Hobbs, D. W., & Taylor, M. G. (2000). Nature of the thaumasite sulfate attack mechanism in field concrete. *Cement and Concrete Research, 30*(4), 529–533.

Hodhod, O. A., & Salama, G. (2014). Simulation of expansion in cement based materials subjected to external sulfate attack. *Ain Shams Engineering Journal, 5*, 7–15.

Hossack, A., & Thomas, M. D. A. (2015). Varying fly ash and slag contents in Portland limestone cement mortars exposed to external sulfates. *Construction and Building Materials, 78*, 333–341.

Hossack, A., Thomas, M. D. A., & Moffatt, E. (2020). Field performance of Portland limestone cement concretes exposed to cold-temperature sulfate solutions. In E. Menéndez & V. Baroghel-Bouny (Eds.), *Proceedings of the Workshop on External Sulfate Attack. Field Aspects and Lab Tests (TESA 2018), 24–25 May 2018, Madrid, Spain* (RILEM Bookseries, Vol. 21, pp. 3–14). RILEM Publications.

Huang, Q., Wang, C., Yang, C., Zhou, L., & Yin, J. (2015). Accelerated sulfate attack on mortars using electrical pulse. *Construction and Building Materials, 95*, 875–881.

Huang, Q., Wang, C., Zeng, Q., Yang, C., Luo, C., & Yang, K. (2016). Deterioration of mortars exposed to sulfate attack under electrical field. *Construction and Building Materials, 117*, 121–128.

Idiart, A. E. (2009). *Coupled analysis of degradation processes in concrete specimens at the meso-level*. Ph.D. thesis, University Politècnica de Catalunya, Barcelona, Spain.

Idiart, A. E., López, C. M., & Carol, I. (2011). Chemo-mechanical analysis of concrete cracking and degradation due to external sulfate attack: A meso-scale model. *Cement and Concrete Composites, 33*(3), 411–423.

Ikumi, T., Cavalaro, H. P. S., Segura, I., de la Fuente, A., & Aguado, A. (2016). Simplified methodology to evaluate the external sulfate attack in concrete structures. *Materials and Design, 89*, 1147–1160.

Irassar, E. F., Bonavetti, V. L., & Gonzalez, M. (2003). Microstructural study of sulphate attack on ordinary and limestone Portland cements at ambient temperature. *Cement and Concrete Research, 33*, 31–41.

IREX/Projet National PerfDub. (2023). *Réaction sulfatique externe (RSE) [External sulfate reaction (ESR)] (in French), in Approche performantielle de la durabilité des ouvrages en béton – De la qualification en laboratoire au suivi d'exécution* (pp. 50–81). Eyrolles.

Jabbour, M. (2019). *Multi-scale study of external sulfate attack in cementitious materials and reinforced concrete structures*. Ph.D. thesis, University of Paris-Est, France.

Jabbour, M., Omikrine-Metalssi, O., Quiertant, M., & Baroghel-Bouny, V. (2019, June 5–7). Physico-mechanical aspects of external sulfate attack: Impact of the cement mix design and the sample shape on the expansion. In *Proceedings of the 9th International Conference on Concrete under severe conditions – Environment and loading (CONSEC'2019)*, Porto Alegre/RS, Brazil.

Khan, M. S. H., Kayali, O., & Troitzsch, U. (2017). Effect of NaOH activation on sulphate resistance of GGBFS and binary blend pastes. *Cement and Concrete Composites, 81*, 49–58.

Khatri, R. P., Sirivivatnanon, V., & Yang, J. L. (1997). Role of permeability in sulfate attack. *Cement and Concrete Research, 27*(8), 1179–1189.

Koch, A., & Steinegger, H. (1960). Ein Schnellprüfverfahren für Zemente auf ihr Verhalten bei Sulfatangriff (A rapid test method for cements to assess their behavior under sulfate attack) (in German). *Zement-Kalk-Gips, 13*(7), 317–324.

Kunther, W., Lothenbach, B., & Scrivener, K. L. (2013). On the relevance of volume increase for the length changes of mortar bars in sulfate solutions. *Cement and Concrete Research, 46*, 23–29.

Lafuma, H. (1929). Théorie de l'expansion des liants hydrauliques [Theory of expansion of hydraulic binders] (in French). *La revue de matériaux de construction et de travaux publics, Part 1, 243*, 441–444. and Part 2, 244 (Jan. 1930), 4–8.

Lawrence, C. D. (1995). Mortar expansions due to delayed ettringite formation effects of curing period and temperature. *Cement and Concrete Research, 25*(4), 903–914.

Lee, S., Hooton, R., Jung, H., Park, D., & Choi, C. (2008a). Effect of limestone filler on the deterioration of mortars and pastes exposed to sulphate solutions at ambient temperatures. *Cement Concrete Research, 38*, 68–76.

Lee, S.-T., Park, D.-W., & Ann, K.-Y. (2008b). Mitigating effect of chloride ions on sulfate attack of cement mortars with or without silica fume. *Canadian Journal of Civil Engineering, 35*, 1210–1220. https://doi.org/10.1139/L08-065

Leemann, A., & Loser, R. (2012). Accelerated sulfate resistance test for concrete-chemical and micro-structural aspects. In G. Ye, K. Van Breugel, W. Sun, & C. Miao (Eds.), *Proceedings of the 2nd International Conference on Microstructural-related Durability of Cementitious Composites, April 11–13, 2012, Amsterdam, The Netherlands* (PRO 83). RILEM Publications, SARL.

Linger, L., Roziere, E., Cussigh, F., Rougeau, P., & Loukili, A. (2014). Concrete equivalent performance concept for durability – An operational guide for the comparative approach. In *Proceedings of the Fib Congress*.

Lipus, K. (2020). New approach for testing SR Cements. In E. Menéndez & V. Baroghel-Bouny (Eds.), *Proceedings of the Workshop on External Sulfate Attack – Field aspects and lab tests (TESA 2018), 24–25 May 2018, Madrid, Spain* (RILEM Bookseries, Vol. 21, pp. 107–120). RILEM Publications.

Lorente, S., Yssorche-Cubaynes, M.-P., & Auger, J. (2011). Sulfate transfer through concrete: Migration and diffusion results. *Cement and Concrete Composites, 33*, 735–741.
Loser, R., & Leemann, A. (2016). An accelerated sulfate resistance test for concrete. *Materials and Structures, 49*, 3445–3457.
Lothenbach, B., & Winnefeld, F. (2006). Thermodynamic modelling of the hydration of Portland cement. *Cement and Concrete Research, 36*, 209–226.
Lothenbach, B., Winnefeld, F., Alder, C., Wieland, E., & Lunk, P. (2007). Effects of temperature on the pore solution, microstructure and hydration products of Portland cement pastes. *Cement and Concrete Research, 37*(4), 483–491.
Lothenbach, B., Bary, B., Le Bescop, P., Schmidt, T., & Leterrier, N. (2010). Sulfate ingress in Portland cement. *Cement and Concrete Research, 40*, 1211–1225.
Ma, X., Çopuroglu, O., Schlangen, E., Han, N., & Xing, F. (2018). Expansion and degradation of cement paste in sodium sulfate solutions. *Construction and Building Materials, 158*, 410–422.
Maes, M., & De Belie, N. (2014). Resistance of concrete and mortar against combined attack of chloride and sodium sulphate. *Cement and Concrete Composites, 53*, 59–72.
Massaad, G. (2016). *Stratégie performantielle d'évaluation de la résistance des matériaux cimentaires à l'attaque sulfatique externe* [*Performance-based strategy for evaluating the resistance of cementitious materials to external sulfate attack*] (in French). Ph.D. thesis, University of Nantes, France.
Massaad, G., Rozière, E., Loukili, A., & Izoret, L. (2015, May 27–29). Nouvelle méthode d'évaluation des attaques sulfatiques externes: analyse des changements microstructuraux à partir de données macroscopiques [New method for assessing external sulfate attacks: Analysis of microstructural changes from macroscopic data] (in French). In *Proceedings of the XXXIII*[èmes] *Rencontres Universitaires de Génie Civil*, Bayonne, France.
Massaad, G., Roziere, E., Loukili, A., & Izoret, L. (2016a). Advanced testing and performance specifications for the cementitious materials under external sulfate attacks. *Construction and Building Materials, 127*, 918–931.
Massaad, G., Roziere, E., Loukili, A., & Izoret, L. (2016b). Advanced external sulfate attack testing and performance specifications: Different sample geometries testing. *Key Engineering Materials, 711*, 1045–1052.
Massaad, G., Rozière, E., Loukili, A., & Izoret, L. (2017). Do the geometry and aggregates size influence external sulfate attack mechanism? *Construction and Building Materials, 157*, 778–789.
Matschei, T., Lothenbach, B., & Glasser, F. P. (2007). The AFm-phase in Portland cement. *Cement and Concrete Research, 37*(2), 118–130.
Mehta, P. K. (1973). Mechanism of expansion associated with ettringite formation. *Cement and Concrete Research, 3*, 1–6.
Mehta, P. K. (1975). Evaluation of sulfate-resisting cements by a new test method. *ACI Materials Journal, 72*(40), 573–575.
Menéndez, E., Garcia–Roves, R., Aldea, B., Ruiz, S., & Baroghel-Bouny, V. (2019). Combination of immersion and semi-immersion tests to evaluate concretes manufactured with sulphate resisting cements. *Journal of Sustainable Cement-Based Material, 8*(6), 337–352.
Menéndez, E., Aldea, B., Formoso, M., Ruiz, S., & de Frutos, J. (2020a). Influence of temperature and aggressive solutions in the formation of thaumasite and ettringite in standard and commercial mortars. In E. Ménendez & V. Baroghel-Bouny (Eds.), *Proceedings of the Workshop on External Sulfate Attack. Field Aspects and Lab Tests (TESA 2018), 24–25 May 2018, Madrid, Spain* (RILEM Bookseries, Vol. 21, pp. 55–70). RILEM Publications.
Menéndez, E., García-Rovés, R., Aldea, B., & Salem, Y. (2020b). Review of the incidence of the sulphate attack in Spain. Evaluation of field concrete cases. In E. Menéndez & V. Baroghel-Bouny (Eds.), *Proceedings of the Workshop on External Sulfate Attack. Field Aspects and Lab Tests (TESA 2018), 24–25 May 2018, Madrid, Spain* (RILEM Bookseries, Vol. 21, pp. 15–32). RILEM Publications.
Messad, S. (2009). *Mise au point d'un essai de vieillissement accéléré de l'attaque sulfatique externe pour l'application du concept de performance équivalente dans le cadre de la norme*

NF EN 206-1 [*Development of an accelerated ageing test of external sulfate attack for the application of the equivalent performance concept within the framework of NF EN 206-1 standard*] (in French). Ph.D. thesis, University of Toulouse, France.

Messad, S., Carcasses, M., Linger, L., & Boutillon, L. (2010). Performance approach using accelerated test method for external sulfate attack. In *Proceedings of the conference Marine environment damage to coastal and historical structures (MEDACHS'10)*, La Rochelle, France.

Meulenyzer, S., Huet, B., Raqen, H., Bégaud, F., Bui, N. N., Vandamme, M., & Pereira, J. M. (2020). A fast testing method for discriminating hardened cement paste reactivity with external sulfate. In E. Menéndez & V. Baroghel-Bouny (Eds.), *Proceedings of the Workshop on External Sulfate Attack – Field aspects and lab tests (TESA 2018), 24–25 May 2018, Madrid, Spain* (RILEM Bookseries, Vol. 21, pp. 121–138). RILEM Publications.

Mittermayr, F., Baldermann, A., Kurta, C., Rinder, T., Klammer, D., Leis, A., Tritthart, J., & Dietzel, M. (2013). Evaporation – A key mechanism for the thaumasite form of sulfate attack. *Cement and Concrete Research, 49*, 55–64.

Mota, B., Matschei, T., & Scrivener, K. (2015). The influence of sodium salts and gypsum on alite hydration. *Cement and Concrete Research, 75*, 53–65.

Mulenga, D. M. (2002). *Zum Sulfatangriff auf Beton und Mörtel einschließlich der Thaumasitbildung* [*Sulfate attack on concrete and mortar, including thaumasite formation*] (in German). Dissertation, Weimar, Germany.

Müllauer, W., Beddoe, R. E., & Heinz, D. (2013). Sulfate attack expansion mechanisms. *Cement and Concrete Research, 52*, 208–215.

Nanocem Website. (2004–2021). https://www.nanocem.org

Nehdi, M. L., Suleiman, A. R., & Soliman, A. M. (2014). Investigation of concrete exposed to dual sulfate attack. *Cement and Concrete Research, 64*, 42–53.

Neunzig, C., Heiermann, T., & Raupach, M. (2018). Determination of the uniaxial tensile strength of concrete with a modified test setup. In V. Mechtcherine, V. Slowik, & P. Kabele (Eds.), *Proceedings of the International Conference on Strain-Hardening Cement-Based Composites (SHCC4), 2017, Dresden, Germany* (RILEM Bookseries, Vol. 15, pp. 316–323). RILEM Publications.

Neville, A. (2004). The confused world of sulfate attack on concrete. *Cement and Concrete Research, 34*(8), 1275–1296.

NF P15-317. (2023, November). *Liants hydrauliques – Ciments pour travaux à la mer* [*Hydraulic binders – Seawater resisting cements*] (in French). AFNOR.

NF P18-459. (2022, August). *Béton – Essai pour béton durci – Essai de porosité et de masse volumique* [*Concrete – Testing hardened concrete – Testing porosity and density*] (in French). AFNOR.

Perfdub Website. (2014–2019). http://www.perfdub.fr/en/

Ping, X., & Beaudoin, J. J. (1992a). Mechanism of sulfate expansion. I. Thermodynamic principles of crystallization pressure. *Cement and Concrete Research, 22*, 631–640.

Ping, X., & Beaudoin, J. J. (1992b). Mechanism of sulfate expansion. II. Validation of thermodynamic theory. *Cement and Concrete Research, 22*, 845–854.

Pipilikaki, P., Papageorgiou, D., Teas, C., Chaniotakis, E., & Katsioti, M. (2008). The effect of temperature on thaumasite formation. *Cement and Concrete Composites, 30*, 964–969.

Planel, D., Sercombe, J., Le Bescop, P., Adenot, F., & Torrenti, J. M. (2006). Long-term performance of cement paste during combined calcium leaching-sulfate attack: Kinetics and size effect. *Cement and Concrete Research, 36*, 137–143.

Ramezanianpour, A. M., & Hooton, R. D. (2013). Sulfate resistance of Portland-limestone cements in combination with supplementary cementitious materials. *Materials and Structures, 46*, 1061–1073.

Revertegat, E., Richet, C., & Gegout, P. (1992). Effect of pH on the durability of cement pastes. *Cement and Concrete Research, 22*, 259–272.

RILEM. (2007). *Proceedings of the International RILEM workshop on performance based evaluation and indicators for concrete durability*, March 19–21, 2006, Madrid, Spain, PRO 47 (V. Baroghel-Bouny, C. Andrade, R. Torrent, & K. Scrivener, Eds.). RILEM Publications.

RILEM. (2013). *Performance of cement-based materials in aggressive aqueous environments* (State-of-the-Art Report, RILEM TC 211-PAE) (M. Alexander, A. Bertron, & N. De Belie, Eds.). RILEM.

RILEM. (2018). Properties of fresh and hardened concrete containing supplementary cementitious materials, state-of-the-art report, RILEM TC 238-SCM, working group 4. In N. De Belie, M. Soutsos, & E. Gruyaert (Eds.), (p. 315). RILEM.

Rozière, E. (2007). *Etude de la durabilité des bétons par une approche performantielle* [*Study of the durability of concrete using a performance-based approach*] (in French). Ph.D. thesis, Ecole Centrale de Nantes, France.

Rozière, E., Loukili, A., El Hachem, R., & Grondin, F. (2009). Durability of concrete exposed to leaching and external sulphate attack. *Cement and Concrete Research, 39*, 1188–1198.

Rozière, E., El Hachem, R., & Loukili, A. (2015). Chapter 5. Construction degradation by external sulfate attacks. In A. Ait-Mokhtar & O. Millet (Eds.), *Structure design and degradation mechanisms in coastal environments*. Wiley-ISTE. ISBN: 978-1-84821-732-4, 50p.

Sahmaran, M., Kasap, O., Duru, K., & Yaman, I. O. (2007). Effects of mix composition and water-cement ratio on the sulfate resistance of blended cements. *Cement and Concrete Composites, 29*, 159–167.

Saillio, M., Baroghel-Bouny, V., & Pradelle, S. (2016, September 12–14). Effect of carbonation and sulphate on chloride ingress in cement pastes and concretes with supplementary cementitious materials. In M. Colombo & M. di Prisco (Eds.), *Proceedings of the 8th International Conference on Concrete under severe conditions – Environment and loading (CONSEC'2016), Lecco, Lake Como, Italy* (pp. 241–248). Trans Tech Publications, Pfaffikon, Switzerland.

Saillio, M., Bertin, M., & Baroghel-Bouny, V. (2019, March 20–22) Coupling effect and durability in cement pastes and concretes with supplementary cementitious materials. In *Proceedings of the International Conference on Sustainable materials, systems and structures (SMSS 2019), Rovinj, Croatia* (Segment: Durability, Monitoring and Repair of Structures, PRO 18, pp. 436–443). RILEM Publications, SARL, Paris.

Santhanam, M., Cohen, M. D., & Olek, J. (2001). Sulfate attack research – Whither now? *Cement and Concrete Research, 31*, 845–851.

Santhanam, M., Cohen, M. D., & Olek, J. (2002a). Mechanism of sulfate attack: A fresh look: Part 1: Summary of experimental results. *Cement and Concrete Research, 32*(6), 915–921.

Santhanam, M., Cohen, M. D., & Olek, J. (2002b). Modeling effect of solution concentration and temperature during sulfate attach on cement mortars. *Cement and Concrete Research, 32*(4), 585–592.

Santhanam, M., Cohen, M. D., & Olek, J. (2003). Effects of gypsum formation on the performance of cement mortars during external sulphate attack. *Cement and Concrete Research, 33*, 325–332.

Santhanam, M., Cohen, M. D., & Olek, J. (2006). Differentiating seawater and groundwater sulfate attack in Portland cement mortars. *Cement and Concrete Research, 36*(12), 2132–2137.

Santillán, L. R., Locati, F., Villagrán-Zaccardi, Y. A., & Zega, C. J. (2020). Long-term sulfate attack on recycled aggregate concrete immersed in sodium sulfate solution for 10 years. *Materiales de Construcción, 70*(337), e212.

Scherer, G. W. (2004). Stress from crystallization of salt. *Cement and Concrete Research, 34*, 1613–1624.

Schmidt, T., Lothenbach, B., Romer, M., Scrivener, K., Rentsch, D., & Figi, R. (2008). A thermodynamic and experimental study of the conditions of thaumasite formation. *Cement and Concrete Research, 38*, 337–349.

Schmidt, T., Lothenbach, B., Romer, M., Neuenschwander, J., & Scrivener, K. (2009). Physical and microstructural aspects of sulfate attack on ordinary and limestone blended Portland cements. *Cement and Concrete Research, 39*, 1111–1121.

Scrivener, K. L., & Taylor, H. F. W. (1993). Delayed ettringite formation: A microstructural and microanalytical study. *Advances in Cement Research, 5*(20), 139–146.

SIA 262/1. (2019, January). *(SN 505 262/1): Concrete structures – Supplementary specifications – Annex D.* SIA.
Skalny, J., & Pierce, J. (1999). Sulfate attack issues. In J. Skalny & J. Marchand (Eds.), *Material science of concrete – Sulfate attack mechanisms* (pp. 49–64). American Ceramic Society.
Soive, A., & Tran, V. Q. (2017). External sulfate attack of cementitious materials: New insights gained through numerical modeling including dissolution/precipitation kinetics and surface complexation. *Cement and Concrete Composites, 83*, 263–272.
Stephens, J. B., & Carrasquillo, R. L. (2000). *Evaluating performance-based test and specifications for sulfate resistance in concrete* (Research Report 1706-3). University of Texas.
Suma, M. F., & Santhanam, M. (2020). Influence of specimen size on the expansion of Portland cement, mortars immersed in sodium sulphate solution. In E. Menéndez & V. Baroghel-Bouny (Eds.), *Proceedings of the workshop on external sulfate attack. Field aspects and lab tests (TESA 2018), 24–25 May 2018, Madrid, Spain* (RILEM Bookseries, Vol. 21, pp. 81–92). RILEM Publications.
Swenson, E. G. (1971). *Concrete in sulphate environments* (CBD-136). Canadian Build Dig.
Tikalsky, P. J., & Carrasquillo, R. L. (1989). *The effect of fly ash on the sulfate resistance of concrete* (Research Report 481-5). University of Texas.
Torres, S. M., Leal, A. F., Vieira, A. A. P., Barbosa, N. P. (2011). Thaumasite form of sulphate attack in a tropical climate weather. In A. Palomo (Ed.), *Proceedings of the XIII International Congress on the Chemistry of Cement* (Paper 901, p. 401). Madrid.
Tumidajski, P. J., & Turc, I. (1995). A rapid test for sulfate ingress into concrete. *Cement and Concrete Research, 25*(5), 924–928.
USBR 4908-86. (1986). *Procedure for length change of hardened concrete exposed to alkali sulfates.* United States Bureau of Reclamation, US Department of the Interior, US Government Printing Office.
Van der Lee, J., De Windt, L., Lagneau, V., & Goblet, P. (2003). Module oriented modeling of reactive transport with HYTEC. *Computational Geosciences, 29*, 265–275.
VDZ Report. (2017). *Beurteilung des SVA-Flachprismenverfahrens zur Bestimmung des Sulfatwiderstands von Zementen und Ableitung eines modifizierten, praxisgerechten Prüfverfahrens [Research results: Evaluation of the SVA flat prism method for determining the sulfate resistance of cements and derivation of a modified, more practical test method]* (in German) (AiF Research Project No 18024 N, Research Report). VDZ.
Wan Kai Tai. (2016). *Feasibility study of modied ring test for sulphate resistance testing, report.* Brunel University.
Whittaker, M., & Black, L. (2015). Current knowledge of external sulphate attack. *Advances in Cement Research, 27*, 1–14.
Wittekindt, W. (1960). Sulfatbeständige Zemente und ihre Prüfung (Sulfate-resistant cements and their testing) (in German). *Zement-Kalk-Gips, 13*(2), 565–572.
XP P18-462. (2022, December). *Essai sur béton durci – Essai accéléré de migration des ions chlorure en régime non-stationnaire – Détermination du coefficient de diffusion apparent des ions chlorure [Testing hardened concrete – Chloride ions migration accelerated test in non-steady-state conditions – Determining the apparent chloride ions diffusion coefficient]* (in French). AFNOR.
Xu, Y. (1997). The influence of sulphates on chloride binding and pore solution chemistry. *Cement and Concrete Research, 27*(12), 1841–1850.
Yang, D. Y., & Guo, R. (2014). Experimental study on modulus and hardness of ettringite. *Experimental Techniques, 38*(1), 6–12.
Yoshida, N. (2019). Sulfate attack on residential concrete foundations in Japan. *Journal of Sustainable Cement-Based Materials, 8*(6), 327–336.
Yu, P., & Kirkpatrick, R. J. (2001). ^{35}Cl NMR relaxation study of cement hydrate suspensions. *Cement and Concrete Research, 31*(10), 1479–1485.
Yu, C., Scrivener, K., & Sun, W. (2012). Comparison between expansion and microstructural changes of mortar under sulfate attack. In G. Ye, K. Van Breugel, W. Sun, & C. Miao (Eds.),

Proceedings of the second international conference on Microstructural-related Durability of Cementitious Composites, 12–13 April 2012, Amsterdam, The Netherlands, PRO 83 (pp. 595–603). RILEM Publications, SARL.

Yu, C., Sun, W., & Scrivener, K. (2013). Mechanism of expansion of mortars immersed in sodium sulfate solutions. *Cement and Concrete Research, 43*, 105–111.

Zhutovsky, S., & Hooton, R. D. (2017). Accelerated testing of cementitious materials for resistance to physical sulfate attack. *Construction and Building Materials, 145*, 98–106.

Zuquan, J., Wei, S., Yunsheng, Z., Jinyang, J., & Jianzhong, L. (2007). Interaction between sulfate and chloride solution attack of concretes with and without fly ash. *Cement and Concrete Research, 37*, 1223–1232.

The manufacturer's authorised representative in the EU is Springer Nature Customer Service Centre GmbH, Europaplatz 3, 69115 Heidelberg, Germany. If you have any concerns regarding our products, please contact ProductSafety@springernature.com

Printed and bound by CPI Group (UK) Ltd, Croydon, CR0 4YY
07/10/2026
02229357-0002